CLEAN & QUIET

THE GUIDE TO ELECTRIC POWERED FLIGHT

BOB ABERLE

edited by FRANK FANELLI

Printed in the United States of America

ISBN 0-924771-58-5
10 9 8 7 6 5 4 3

Dedication

This book is dedicated to three people, all of whom greatly influenced its publication.

First to my wife, Irene, for giving me the space and understanding to let me pursue my hobby interests over all these years, and most of all for her constant encouragement which included the writing of this book.

Second to Bob Kopski, electric columnist of *Model Aviation* magazine, the man who has spent years preaching and teaching the virtues of electric powered flight. He made me a believer for sure!

Finally to my closest friend, flying partner and business partner, Tom Hunt. To Tom, nothing is impossible. We have interacted on a daily basis for many years. Anything asked of Tom is done in a day's time, without exception. There is no such thing as "can't be done" in his vocabulary. I'm proud of Tom for everything he has accomplished to date and for what I know will come in the future. Just having a little of his technical and flying skills rub off on me has been a wonderful experience. Thanks, Tom!

Acknowledgement

Any undertaking such as this requires help from many individuals. My "bride" of 32 years, Irene, read and made suggestions and corrections throughout the text.

A special thanks goes to my neighbor, Bert Liebowitz, who provided his comments from the perspective of the newcomer to the hobby. To my friend of many years, Larry Sribnick, founder and President of SR Batteries Inc. thanks are due for his industry viewpoint of the manuscript. And to Don Bousquet, that wonderful cartoonist and modeler, who first gave me his comments based on being both an author and hobbyist. Don was gracious enough to do a series of his cartoons expressly for this book.

Finally a special thanks to my buddy and Associate Editor of *Flying Models* magazine, Frank Fanelli, for his patience and excellent skills in editing and organizing this book into the final product. This was truly a partnership of author and editor, equally sharing. The end result was both rewarding and fun!

Contents

1. Introduction	1
2. Advantages	7
3. Flight Systems	13
4. Motors	17
5. Props	25
6. Batteries & Chargers	29
7. Speed Controllers	41
8. Connectors & Wires	49
9. Fuses, Switches, Charging	55
10. Radio Systems	59
11. Selecting Powertrains	63
12. Suitable Aircraft	69
13. Aircraft Selection	75
14. Flying Electrics	81
15. Knowing Electrics More	85
16. Future of Electrics	89
Appendix	91

Foreword

Electric power can provide the quiet and ecologically friendly way to enjoy the hobby and sport of flying model aircraft. The purpose of this book is to introduce you to modern electric power systems, to tell you the advantages of this kind of power and how easily it can be adapted to the flight of model aircraft. It takes you from the beginning up till the present, with sufficient depth to get you from the kit box and individual components to the flying field. Quiet and clean electric power is here now for the benefit of all model aviation enthusiasts.

About the author

Bob Aberle is a retired aerospace engineering manager who has devoted 45 years to the hobby of model aviation. Starting with free flight models in 1950, Bob quickly got involved in R/C in its early stages. For the past 20 years, on a part time basis, he has been first Contributing Editor and later, Technical Editor of *Flying Models* magazine. He has been responsible for the publication of over 30 original R/C designs covering all facets of the hobby and for some 2,000 pages of articles covering product reviews, how-to's, contest reports and the like.

In the early eighties, Bob chaired the prestigious AMA R/C Frequency Committee that was successful in obtaining from the FCC the present 80 R/C channels that we use today to operate our models. For that effort he was appointed AMA Fellow in 1983, receiving also the AMA's Distinguished Service Award.

Additionally Bob was awarded the Howard McEntee Memorial Award in 1982 and in 1993 he was inducted into the Vintage R/C Society Hall of Fame.

Bob Aberle flew his first electric powered model aircraft in 1979. Today it is essentially the only form of power that he employs as a modeler. As a point of current information, Bob attended and competed in the very first National Electric Championship Meet to be held at the new AMA Headquarters Facility in Muncie, Indiana in June 1995. He won a third place in both the Class-A and the Class B Old Timer Events.

electric flight: an introduction

There are probably many model aircraft enthusiasts today who feel that electric powered flight is a new technology that is still in its infancy. Many of these same folks continue to take a "wait and see" attitude, presumably anticipating more technical progress in the future. Well, there is no need to wait any longer: the technology is here now and it can provide you with considerable enjoyment and satisfaction if you are willing to take up the challenge.

Lozier's Experimental Gas-Electric Model

HERE, fellows, is a unique monoplane with which you can have a lot of fun. And since it is adaptable in the matter of power, you can fly it with your regular gas model motor or an electrical engine. The motor in the original electric model came from a discarded automobile horn and was "juiced" by two C-batteries. It gave excellent results!

Since the details of the job will be governed by your power plant, the drawings below are to be taken as a general guide and modified to suit your needs. For a gas engine, the regular mount can be used in connection with the struts shown. All supporting struts are reinforced with bamboo and are screwed firmly to the fuselage cross-members. Be careful there is no splitting.

An electric motor should be bolted to two pine cross-pieces and then attached to the supporting struts. Dots indicate the positive wiring and dot-dashes show the negative. The positive leads from the motor to a tinfoil fuse and then to the switch. The negative runs from the motor to the battery and then to the switch. Length of flight is controlled, if desired, by number of tinfoil layers on fuse. Chief objection to the electric motor is expense of batteries.

Balsa 3/16" sq. is used for all fuselage members. Hard balsa 1/16" sq. is used for built-up, tapered wing spars, and it likewise is used on all parts drawn in heavy black.

SCALE ⅛" = 1"

EXPERIMENTAL MONOPLANE FOR GASOLINE OR ELECTRIC MOTOR

A FLYING ACES Magazine Plan

[47]

This reprint from page 47 of the 1938 Flying Aces magazine shows Herb Lozier's Experimental Gas or Electric model. It is believed to be the first electric powered model, although there was no proof that it actually flew.

PHOTO: BOB ABERLE

What would probably surprise many modelers is the fact that the first record of electric powered flight dates back to a report that appeared in the October, 1909 issue of *Model Engineer*, in which an electric powered freeflight model was documented. That claim was never substantiated, but I did find a reproduction of the drawing in a book titled *Electric Flight*, by Dave Day, published in 1983 by Argus Books Ltd. of Great Britain. Keep in mind that 1909 wasn't much beyond the Wright Brothers first manned flight.

It was only by accident, recently, that a reference was found to an electric pow-

This model is a replica of Col. H.J. Taplin's Radio Queen. On June 30, 1957 Taplin's model, with an Emerson 300Z 24-volt motor, flew the first recorded flight of an electric powered model. Dave Durnford built this replica of the 84-inch span model and put an Astro Flight 40 geared motor in it and twenty 1400 mAh cells for the battery pack.

PHOTO: DAVE DURNFORD

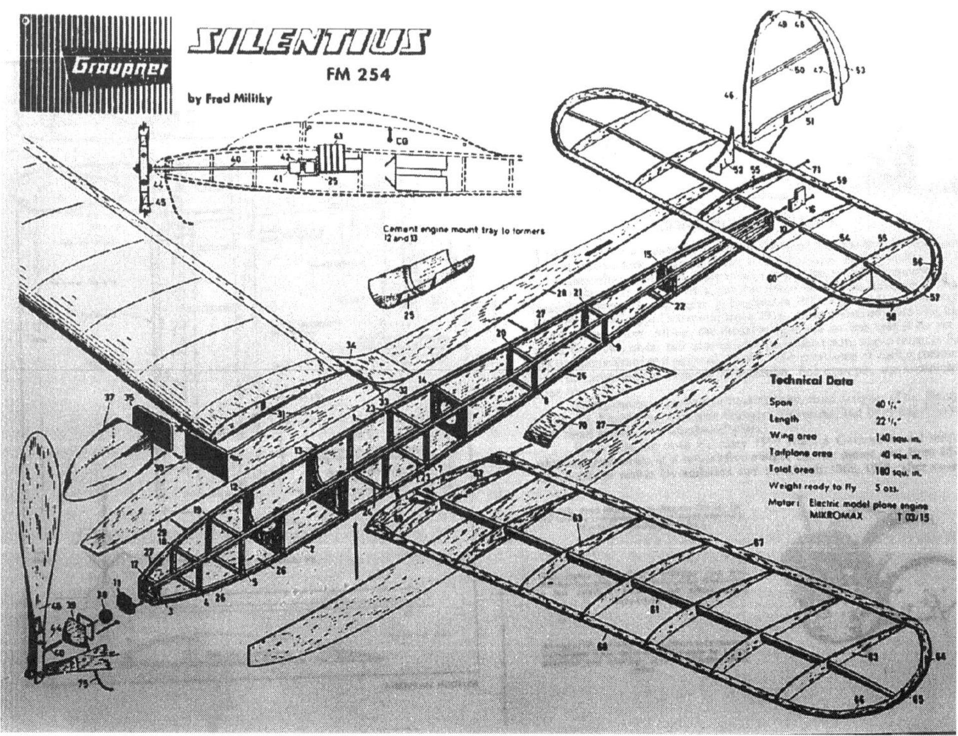

Fred Militky of Germany designed the Silentius strictly for electric power. The model made its first flight in October 1957, and in 1960 became the first commercially available kit exclusively for electric power. The Graupner Compnay of Germany manufactured the kit.

PHOTO: DAVE DURNFORD

ered free-flight model design that appeared in 1938. The specific reference was to an October 1938 issue of *Flying Aces* magazine on page 47. A fellow named Herb Lozier converted an automobile motorized horn into an electric powerplant, and installed it in a model that looked very much like a WW II troop glider. Again there was no proof of flight, but the technical facts presented indicate that actual flight was certainly possible with this design. Remember, the two-cycle gasoline fueled model engine was only developed in the early thirties.

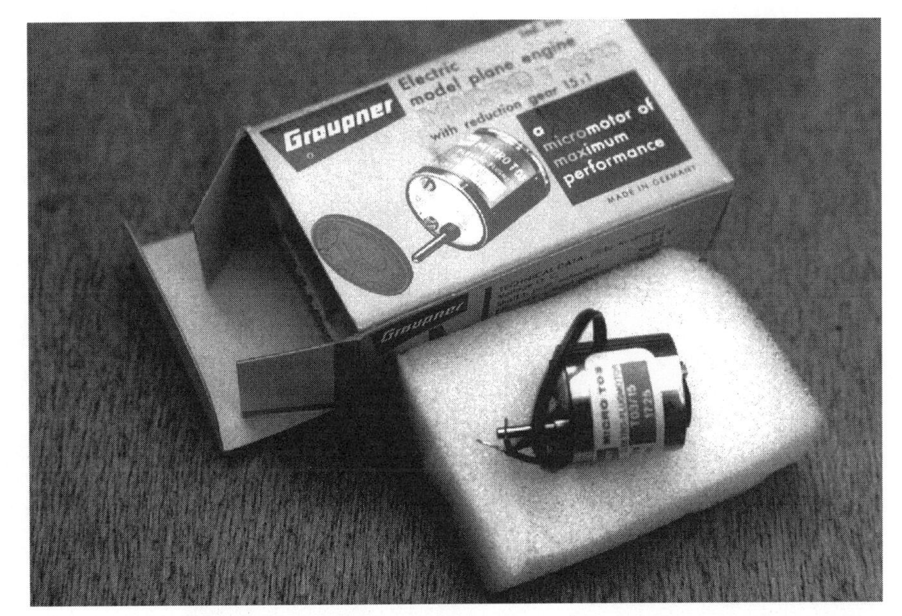

The Graupner Micro-Max TO3/15 motor was used to power the Silentius. It was a Faulhaber motor incorporating an integral 15:1 gear drive.

PHOTO: DAVE DURNFORD

In June 1957, Col. H. J. Taplin (Great Britain) made the first officially recorded flight of an electric powered radio control model. The model, called the Radio Queen, had an all-up weight of eight pounds. It was powered by 25 silver-zinc cells and a war surplus Emerson 24-volt electric motor.

The person given the most credit for getting the electric powered flight movement going is the late Fred Militky of Germany. His electric freeflight design, the Silentius, made its first flight in October 1957. The little model, with 180 square inches of wing area, weighed a total of 5 ounces ready to fly, with both motor and battery. This design was later kitted (1960) by the Graupner Company of Germany and became the first commercial electric powered aircraft kit in the world.

It was in the early sixties that the nickel-cadmium (Ni-Cd) rechargeable battery came on the market. Here at last the electric powered model aircraft flyer had a reasonably lightweight cell, with small size, considerable capacity, and the ability to be recharged in a relatively short period of time.

Using this new battery technology, Bob and Roland Boucher, of California, founded the business known to this day as Astro Flight Inc. Bob's long time development and production of cobalt type electric motors, combined with Ni-Cd battery technology, got us to where we are today in electric powered flight.

The Silentius was a a freeflight model, not radio controlled. This replica of the original Graupner model uses the same Graupner TO3/15 geared motor as the original. It does, however, have a modern 4-cell 50 mAh Sanyo nickel cadmium battery pack. It weighs 102 grams. The original used salt water activated "one-shot" cells.

PHOTO: DAVE DURNFORD

Current state-of-the art technology makes electric powered flight practical, be it sport, pattern, competition, sailplanes, helicopters, old timers, racers, etc. It is as easy today to fly "electrics" as it is a fueled (wet, as we call them!) model. The equipment is all there. So is the overall technology. Best of all, you don't have to be an electrical engineer to fly model aircraft by electrical power. If you can install a radio control system, you can also easily install an electric power system in your model. That will be the basic theme throughout this book.

I felt it necessary, very early in this book, to explain exactly what one should expect from electric powered flight. Pointing out the advantages is one aspect, but the bottom line still gets down to just how well you can expect your model aircraft to perform when it is powered by an electric motor. Many modelers are under the mistaken impression that electric powered model aircraft are only capable of flying for very short periods of time, which then must be followed by long periods of battery recharging. The final result being that you are on the ground much more than in the air. Believe me, that is far from the truth.

In general, everything you can do with a wet fueled model aircraft can be duplicated in an electric powered one. The exact degree they can be duplicated will rely primarily on your gaining some experience in electric power and the ability to manage it properly.

I'm sure if you are already flying with

Flying electrics does require more planning than that required with glow powered models. Bob Aberle's Aeronca L-3 Defender can fly for up to nine minutes with its five 1250 mAh cells. Battery capacity and gearing determine the length of flight with electrics, and it is usually a compromise to determine whether you want speed or duration. His model weighs 33 ounces, and uses an 035 Astro Flight geared motor.

PHOTO: BOB ABERLE

"wet" power, you may have seen a modeler friend show up at the field with an "electric." There is generally a lot of initial curiosity when that happens. More often than not, when the model is flown the results may have proved disappointing.

With the gas powered models, if you want to obtain more flying time, you add a bigger fuel tank. With electrics it isn't quite that easy. Adding flight time to an electric powered model means also adding weight in the form of higher capacity batteries. If you want your gas powered model to fly faster, you might consider a new prop or higher nitro content fuel, neither of which will add weight to your model.

Flying faster on electric power generally involves going to more battery cells which increases the voltage and again the weight. On the plus side for electrics is the distinction between direct motor drive for speed and short motor durations, as opposed to gear or belt reduction drives which provide higher thrust at lower airspeed and longer duration flying time. The proper choice of direct or gear drive is what is makes the difference.

Over the past ten years, electric powered flight has become more practical and even more comparable to gas power. This is primarily due to advances in power management brought about by some dedicated and determined modelers. By carefully selecting the size of the electric motor, whether it should be direct drive or gear reduction drive, the diameter and pitch of the propeller, the number of battery cells (to determine the voltage), the capacity (to establish the desired flying time) and the proper techniques for fully charging the batteries, it is possible to end up with that roughly comparable performance and at the same time gain all the benefits of electric power.

I can consistently obtain 8- to 10-minute flights on a sport scale Aeronca L-3 Defender, powered by a geared 035 motor, where the total weight of the model is just 33 ounces. I also have a pattern model which was converted from glow (.45 size) to electric power and, again, with proper management (making the right choices) I easily obtain 6- to 7-minute flights where I can perform many of the regular pattern maneuvers.

Comparable performance with electric powered flight is a reality right now. It is achieved by a combination of understanding and good power management. This can be easily learned by the average skilled modeler. On-going development of new motor, battery, and speed control technology is only going to make the future of electric powered flight better and better!

advantages

All of this field equipment soon becomes a necessity when flying models with glow fuel engines. You end up needing a glow driver panel, 12-volt battery, starter motor, fuel, fuel pump, and glow plug connector.

PHOTO: BOB ABERLE

Right from the start I can tell you there are far more advantages to electric power than there are disadvantages. In fact I won't even use the word disadvantages, but simply say there are a few "considerations" to keep in mind.

Probably the most important aspect of electric power is the fact that it is very quiet. About the only sound emitted from the model will be propeller wind noise and the actual wind passing over the model's surfaces. Electrics are so quiet I have often resorted to flying at sunrise, while traveling to my place of business. I was able to set up at a vacant athletic field, put in a few flights in the calm morning air and then proceed to work. Never once in following this routine over a period of years did I ever have a single person take notice of what I was doing.

Many of you who have already experienced model aircraft flight using glow fuel (wet) power can attest to the fact that these engines can be a real problem to start in cold winter weather. The engines don't like the cold and neither does the pilot for that matter.

One specific difficulty in flying glow engine models in cold weather is starting them. The methanol-based glow fuel doesn't atomize as easily in colder weather, and results in much more difficult starts, as Bert Leibowitz experiences with his model in 40° weather.

PHOTO: BOB ABERLE

Consider the alternative with electric power; you sit in your car with the engine and heater running while the battery pack is being charged. When the battery is ready, you get out of the car, turn on the radio, arm the electric motor system and launch your model, all in a matter of a few moments. Electric motors will never hesitate or refuse to start in cold weather.

Another real fact and advantage of electric power is that very little vibration is generated while running the motor. As a result the various bolts, screws, nuts, etc. don't seem to work themselves loose. Radio equipment gets better treatment. For the most part less vibration protection (rubber padding) is necessary to protect your equipment.

Since no fuel is employed, your model does not need any special fuel resistant paints or covering material. Trim tapes obtained at local art stores can be applied, without any need for protective overcoats of sealers and the like. And, of course, when you are done flying there is no mess to clean off, since there is no residual fuel (no exhaust) and nothing to rub off on the upholstery of your automobile.

Another advantage of using no fuel is that it is "kind to our environment." We have a serious water supply problem where I live on Long Island. Dumping any kind of chemical into the ground, no matter what the quantity, is a criminal offense. When our local electric club applied for the part-time use of a municipal park activity field, we were initially rejected. We thought at first it was the

Electrics, thanks to their negligible noise and absence of toxic fuels, are welcome in environmentally sensitive areas.

PHOTO: BOB ABERLE

legacy of model engine noise. To our our surprise the rejection was based on the concern that modelers would spill their residual fuels on the ground after flying.

A subsequent series of flight demonstrations, using electric power, easily won us approval to use this first field. Later we were granted the use of an even larger field, which we use to this day. So both advantages—the absence of noise and the absence of "wet" fuel—helped us in this regard.

Still another advantage of electric power is the very real possibility of flying small, lightweight models indoors during the winter months. Many of us who live in the cold winter climates are forced into a "hibernation period," roughly from January through the end of March. Electric powered freeflight and even R/C models are now being flown indoors in sports arenas, auditoriums, gymnasiums, and the like. This is becoming so popular in the Northeast that regular indoor contests are being scheduled throughout the winter. Electric powered R/C models of less than several ounces gross weight are now in the realm of current technology.

With all of the advantages just presented, is there anything that should be "considered" with regard to electric flight? Well the first consideration, of course, is weight. A battery pack will most likely weigh more than a filled fuel tank for a given model size. But the weight differences can be managed properly as you will discover later in this book.

As pointed out, electric motors start up very easily; just turn on a switch and they run. You don't even have to flip the propeller, which is a great finger saver. However, that same feature can cause accidents due to accidental start-ups. I have heard of several modelers who, without thinking, attached a battery pack to an electric motor which had a prop attached to it. The resulting instant start caused the motor to literally run across the work bench reducing everything in its path to pieces and could have also done bodily harm.

You must take precautions at home, enroute to the field, and while at the field, so as not to have any "surprises." Generally speaking, your motor battery

There are four 05 electric motors on Art Thoms' magnificent Boeing 314 Pan Am Clipper. If these were glow engines, the diffculty in starting and adjusting all four engines for consistent runs would be very difficult compared to electric which involves just flipping a switch.

PHOTO: BOB ABERLE

should always be left disconnected except for the time when you are actually flying, even if other safety features, like arming switches and fuses, etc., are used. More about these will be explained later.

Battery charging is another of the "considerations." It is an item that will initially take some practice in developing the proper and safe skills. The charger systems today are becoming very sophisticated, to say the least. A good understanding of how to charge, use, and store your batteries is worth your time to learn properly.

The final consideration is the actual cost of flying with electric power. Can it cost more than "wet" power? This is very important and so part of this chapter is devoted to this subject. All in all, I think

Taking off from water requires more power than a comparable take-off from land. Art Thoms' 114-inch span Custom Privateer successfully takes off from water, without any special techniques or tricks. It weighs 10.9 pounds, and uses an Astro Flight geared 40 with twenty-one 1400 mAh cells.

PHOTO: BOB ABERLE

Sometimes, electrics allow better scale possibilities like Tom Hunt's MODELAIR-TECH Pucara, a scale model of the Argentinian turboprop fighter. Real turboprops have extremely narrow engine cowls becuase of their turbine engines. Trying to fit comparable glow engines with a scale look, with their tall cylinder heads would probably be impossible.

PHOTO: BOB ABERLE

you will agree that the benefits of electric powered flight far exceed the few disadvantages or "considerations" as I call them.

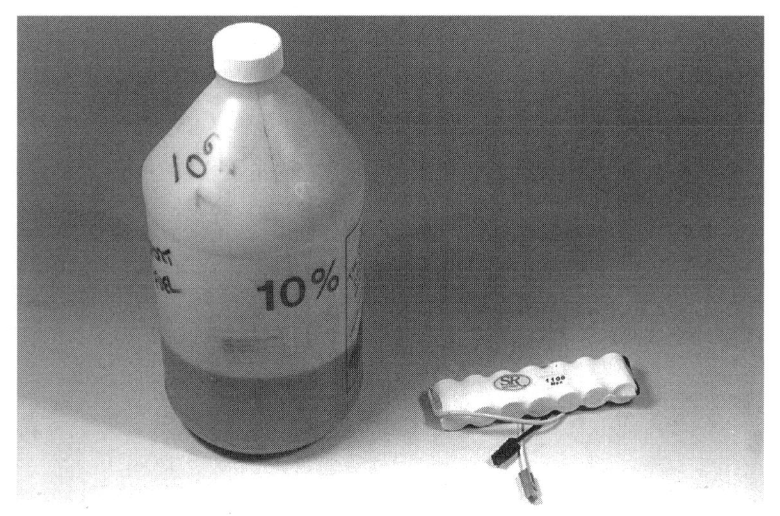

While a gallon of glow fuel costs about half of a 7-cell 1100 mAh nickel cadmium battery pack, the fuel is expended over two hours flying time. The battery can be recharged hundreds of time which makes it far more cost effective than the fuel.

PHOTO: BOB ABERLE

There has been a myth around for years to the effect that electric powered flight costs considerably more than a comparable glow fuel powered model. I thought it was about time that I challenge that myth in the interest of promoting electric flight. In trying to make a meaningful cost analysis between the two forms of power you must make a lot of assumptions and still try to keep the basics on a one-to-one basis.

What I thought would be helpful would be to use a basic trainer kit, available for both glow and electric power, for an honest comparison. The design selected is the popular Perfect Trainer (or PT) by Tower Hobbies Inc. I chose their PT-20 for which I selected an OS .25 FP glow engine to go up against the PT-Electric Trainer powered by a Mabuchi 550 ferrite motor, with a Master Airscrew gear drive. The PT-Electric Trainer is $10.00 more. However, the combined electric motor/gear drive/battery just about equals the more expensive glow engine, so at this point in the analysis, it is a wash.

Next, you will need a radio control system. The same system can be used for either glow or electric power application. For my analysis I chose a Futaba Conquest 4NBF R/C system with three S-148 servos to hold the price down. The three servos can be used on a glow engine model to operate the rudder, elevator and engine throttle. If you choose to add aileron control at a later time, you will need to purchase a fourth servo. On the other hand, with electric power the ON/OFF relay switch or motor speed control plugs directly into the R/C receiver, so a throttle servo isn't necessary. which means that you will be able to fly with aileron, rudder, elevator and motor control (or throttle) using just the three originally purchased servos.

This sounds like electric power in this sense is less expensive. For our purpose in this analysis let's say it is again a wash. Motor controls can vary from $25.00 up to easily $100.00. That choice alone can tilt the scale in either direction.

On the glow engine side you will initially need a group of accessories which include an electric starter, a 12-volt battery and charger, power panel with cables to help light up the glow plug and a

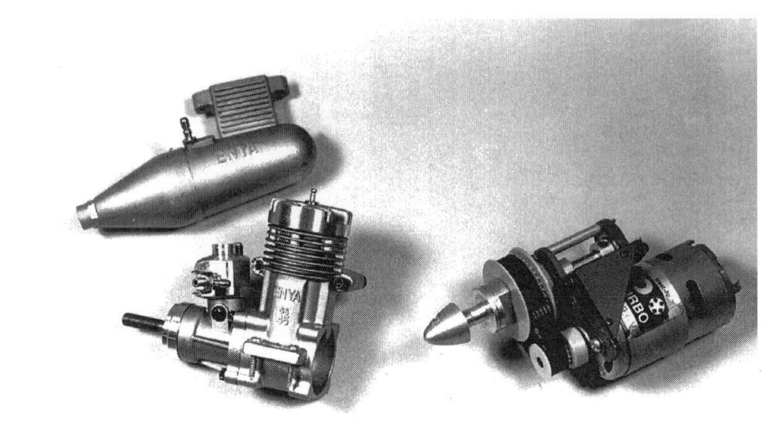

Cost of a modern glow fuel engine and muffler is about equal to a large, comparable electric motor with a belt drive.

PHOTO: BOB ABERLE

three years of constant use (discharging in flight and recharging on the ground). Many active electric flyers will want to own more than one battery pack in order to keep one on charge while the other is being flown. That obviously costs more!

Other expenditures to be considered for electric powered flying include owning a good soldering iron, and having available such items as heavy duty wire, special electrical connectors, fuses and switches.

When I had completed my list and tallied everything on both sides I found that my typical glow engine powered trainer would cost me approximately $470.00, while the equivalent size electric trainer came to a total of $540.00 or roughly 15% more. Keep in mind that many of these costs are what we would classify as "non-recurring." You buy the item once and then use it from plane to plane. Items such as the radio system, glow engine, electric motor and battery charger, can be re-used time after time. You will agree that the cost difference is almost negligible. I hope I convinced you to give electric powered flight a try.

fuel pump to help fill and drain the tank.

The combined cost of those items approximately equals the cost of a good peak detection battery charger to enable you to rapidly charge your motor battery. Again, the cost is comparatively equal, unless you opt for a very sophisticated computer controlled charger. Note: the glow engine supporting equipment physically takes up a lot more room in your vehicle than does the battery charger, which in most cases is powered by your automobile battery.

Building and finishing the model will take the same items such a cement, covering material and an iron to apply the covering. Fuel is a recurring cost or an expendable. At $10-$12 per gallon, you should still expect to get about two hours of flight time on each gallon, possibly more. Motor batteries, if treated properly, could conceivably give you two or

flight systems

The first two chapters encouraged you to try electric powered flight. From this point on you will gradually be introduced into electric powered flight, first as a system and then by the individual components that make up that system.

To sustain flight by electric power you need three major items: an electric motor to turn the propeller, a battery to supply the electric energy to operate the motor and, finally, a switch or control which will allow you to turn the motor on and off or even vary the speed (similar to a throttle) via the radio control system.

In its simplest form you could conceivably wire up an electric powered flight system as shown in **Figure 3-1**. The connections can be permanent (soldered together) without the use of connectors. The switch shown in the block diagram could take on several forms depending on the power level involved. For very low power systems, such as the 35- and 50-watt motors operating on three and four battery cells, a simple toggle switch or lever type micro switch (both obtainable from Radio Shack stores) can be used. Either of these switches may be

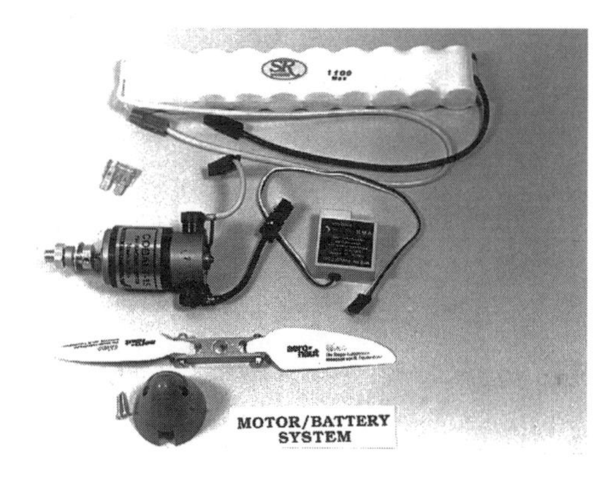

These are the components of a basic electric flight system, with a Robbe relay control. At the top is a 10-cell 1100 SR battery. The motor is an Astro Flight cobalt. The prop is a folding Aeronaut.

PHOTO: BOB ABERLE

Figure 3-1

Block diagram of a basic electric powered flight system using a switch or relay control.

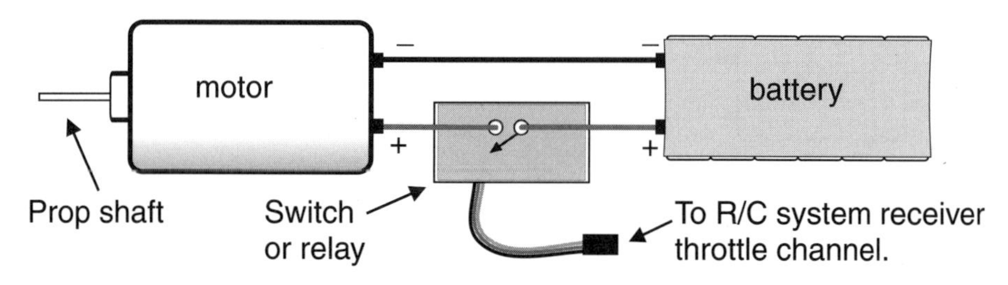

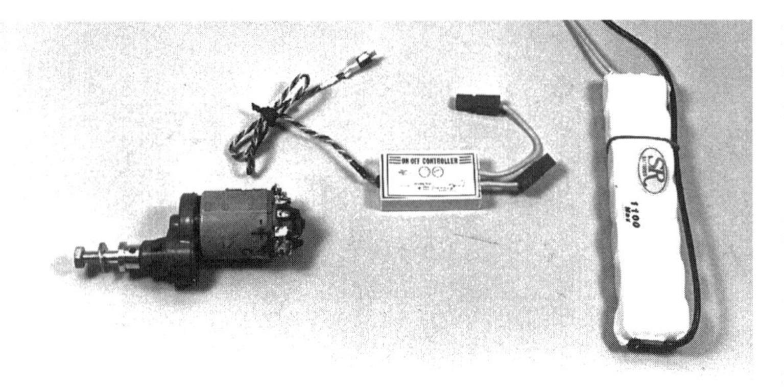

This is a more typical sport flight system setup. The motor is a simple ferrite type, with a Master Airscrew gear drive, a very basic High Sky On/Off controller and a 7-cell battery.

PHOTO: BOB ABERLE

motor is strictly on or off (that's full power or no power). However, for small electric powered sport models this approach is sufficient.

A more popular type of on/off switch control can be obtained from a relay type unit. These devices actually take the place of an R/C servo. The relay contacts can handle loads of up to approximately 25 to 30 amps which gets you well into the general sport/competition area of electric power (upwards of 300 watt motors). Relay

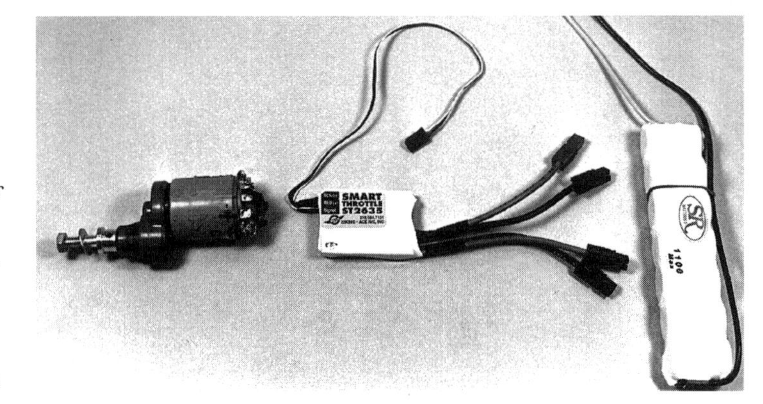

Here's the same ferrite motor, gear drive, and battery as the setup above. However, this system now has an ACE Smart Throttle solid state speed control in place of the High Sky On/Off relay. The Smart Throttle allows proportional (variable) throttle control of the motor.

PHOTO: BOB ABERLE

operated by a small servo that is plugged into the throttle channel of the R/C receiver.

Generally speaking current levels up to 15 amps can be handled in this manner. Keep in mind that the control of the

Figure 3-2 Block diagram of an electric powered flight system using a speed controller.

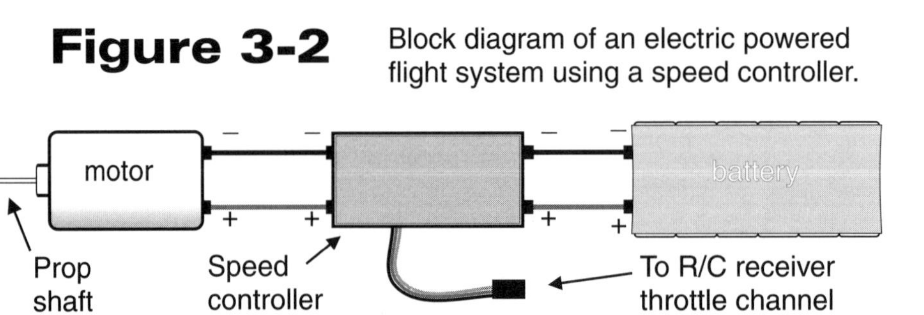

switches will be covered in depth later in **Chapter 7** when the various motor controls are discussed. Again these type switches offer strictly on or off control of the motor. They are not throttle controls.

It is more common today in electric powered flight to see solid state, proportional speed controls used in place of simple switches and relays. These modern controls offer a complete speed range of the electric motor from idle all the way to full power. The amount of power will be in proportion to the movement of the transmitter control stick as it would with "wet" power. **Figure 3-2** shows a block diagram of a basic speed control device inserted between the electric motor and the battery.

You will notice in these diagrams and the ones that follow that there is always a reference to polarity, such as a "+" sign for a positive terminal connection or a "–" sign for the negative connection. General wiring convention also uses red colored wire to denote positive "+" polarity and black colored wire to denote negative polarity "–." In the case of SR Batteries Inc. the popular battery supplier, yellow denotes positive and black negative.

Be sure to always take heed of these "+" and "–" polarity signs when doing any electrical wiring. Crossing connections can, as a minimum, cause your motor to run in the wrong direction. More often, wrong connections can ruin a bat-

On/off motor controls can be as simple as this setup. A servo, hooked to the throttle channel, uses its output arm to trigger a small on/off micro switch connected to the motor.

PHOTO: BOB ABERLE

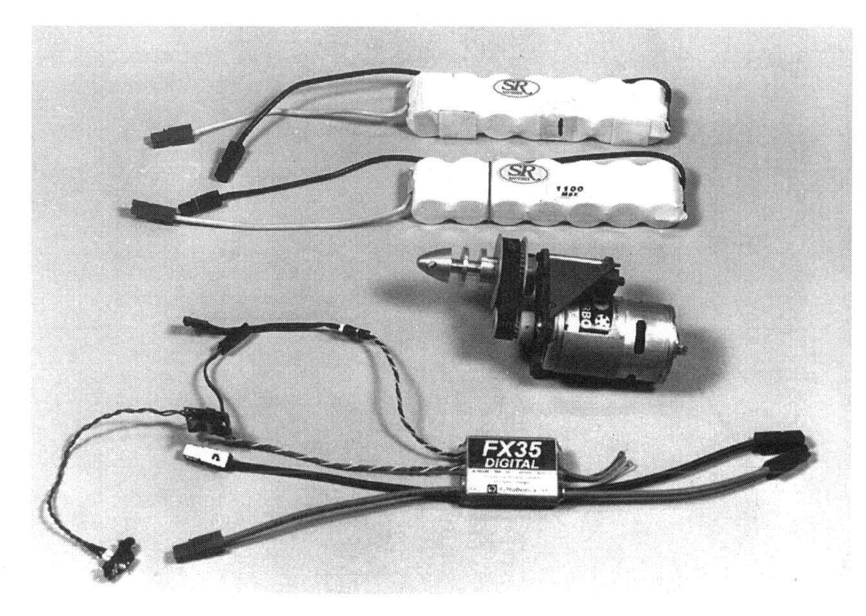

Systems can get more complex, like this one. At the top are two individual 7-cell batteries that will be connected together to form a single 14-cell, 16.8 volt battery to power the Graupner Speed 700 ferrite motor in a belt drive. Below it is an AI-Robotics FX-35D micro-computer solid state speed control.

PHOTO: BOB ABERLE

tery pack or an expensive speed control.

On every motor speed control there will usually be pairs of cables going to both the motor and the battery. There will also be a three-wire cable with a connector which should be plugged into the throttle port (connector) on your R/C receiver. It is important when you buy your speed control that you stipulate the brand of your radio control system so that the cable and connector will be compatible.

In later chapters you will see this basic electric powered flight system expanded to include such practical items as connectors, a master arming switch, a battery charging jack, and selecting the proper size (gauge) wire for your particular system requirements. All of these items must be selected with care to avoid unneccessary power losses in the system. By avoiding losses you get the most power to your motor which, in turn, will provide your best possible flight performance.

motors

Right from the start you should know that a powerplant used for electric flight is called a "motor," not an "engine." Since this is an introductory level book, the intention is to get you started in electric power with the correct choice of motor for your application. It is not my intention to swamp you with technical graphs, curves, specifications, etc., which I'm sure would tend to confuse you more than educate you.

These Graupner Speed 500s are the least expensive ferrite type. From L to R they are the ball-bearing 7.2V Speed 500, a plain bearing 7.2V Speed 500, and a Speed 540 ball bearing.

PHOTO: BOB ABERLE

There are basically three types of electric motors presently on the market, any of which would be suitable to power a model aircraft in flight. These motors are identified as the ferrite, cobalt, and brushless DC types. There is also a fourth type which is starting to gain popularity, which uses a neodymium/iron/boron magnet material (known simply as "neodym" for short).

Neodymium magnet motors are now becoming popular. SR Batteries line of neodymium motors are called the Max series, and will be offered for direct drive, or gear drive.

PHOTO: BOB ABERLE

From an application standpoint, ferrite motors are intended for general sport use and are probably the least efficient. Cobalt and neodym motors are reliable, are long lived, efficient, and are good for both sport and competition flying. The brushless motor offers the longest life, is very efficient, and is an excellent choice for competition flying.

Ferrite type motors are the least expensive, followed by the samarium cobalt (cobalt) and neodym types, and finally the newer variety of brushless DC motors. Prices start as low as $10.00 and go upwards to $500.00 for the largest motors.

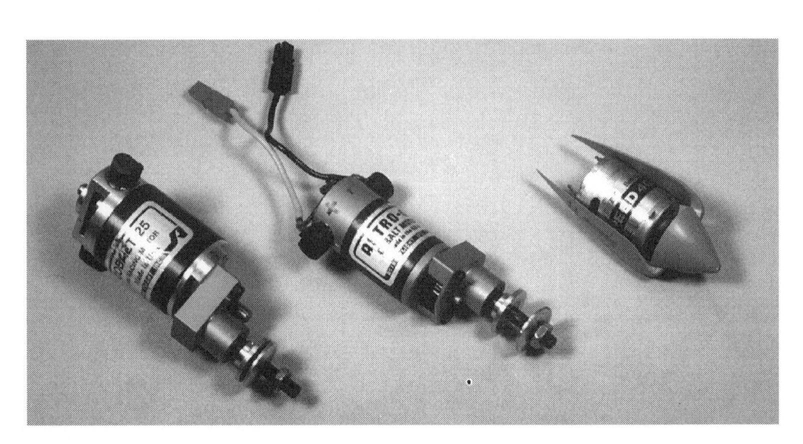

Samarium cobalt motors have been popular in competition, and the Astro Flight cobalts, like the geared 25 at left, and the geared 05, at the right, are among the most popular. At the right is a direct drive Speed 400 ferrite motor shown for comparison.

PHOTO: BOB ABERLE

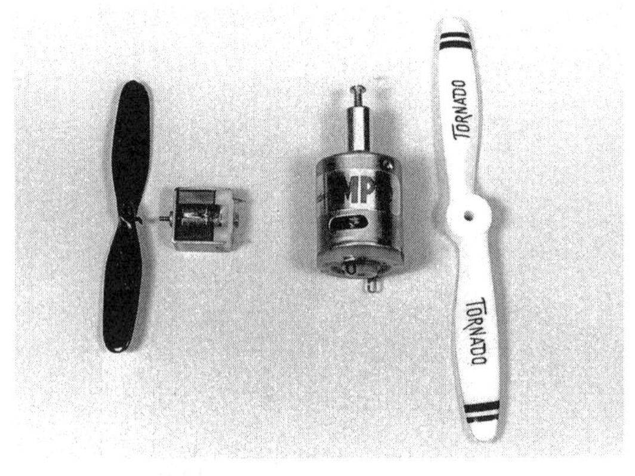

Electric motors can be quite basic, like the tiny HiLine Mini-4 at the left, which runs on only 2–3 cells for freeflight models. The larger IMP-30 at the right runs on three cells.

PHOTO: BOB ABERLE

Generally speaking the ferrite motors are likely to have the shortest life span. The majority of these motors are sold under a variety of brand names, and are totally sealed in a can so to speak. In fact, the name "can motor" seems to have stuck over the years.

Can motors can't be rebuilt or maintained in any way. Therefore, they are essentially disposable. When the brushes get worn, the motor is discarded in favor of a new one.

There are exceptions to this, of course, since some ferrite motors are available today with ball bearing supported armature shafts and with replaceable brush assemblies. Again, the price of these motors makes them very attractive. The ferrite motors as a class are lighter in weight than all the other motor types.

Samarium cobalt motors (or simply "cobalt"), as pioneered in the early '70s by Astro Flight Inc., have become one of the most popular motors for use in electric flight. These are well-built units that have ball-bearing supported shafts and brush assemblies which can easily be replaced. In fact, their gold-colored anodized cases along with the brush assemblies that protrude on either side of the motor casing make the Astro Flights units very recognizable.

In recent years, many new companies have started to manufacture samarium cobalt motors. These motors are long lasting, and come in a wide variety of sizes to handle everything from tiny models of only 25-ounce gross weight, up to ¼ scale aircraft with weights well in excess of 10 pounds. Many of the competitive flyers use cobalt motors for scale, speed, and duration events.

The neodym magnet motors are in this same category. They cost approximately the same as cobalt types and offer most of the advantages. Cobalt and neodym motors generally weigh more than comparable ferrite motors.

The last category of motor, the brushless DC, is rclativcly ncw on thc hobby market. The technology for brushless motors has certainly been around for a long time. It was several years ago that Aveox Inc. introduced a series of brushless DC motors. As the name implies, these motors have no brushes, so there is nothing to wear out. Long term reliability is a key feature of this type motor, along with the claim of improved overall efficiency.

The cost is definitely high right now and the technology does require the use of a special, dedicated speed controller which is both expensive and heavy. However, this technology for model aircraft applications is still quite new and I suspect that weight and cost reduction improvements will only be a matter of time.

If I had to put a price tag on these categories of motors it would be something like this for motors of comparable power output: ferrite motors are in the range of \$10.00 to \$80.00; cobalt and neodym types can run \$60.00 to \$300.00; and the brushless DC motors are in the \$300.00 to \$500.00 range, including their special companion speed controller.

Before you say, "I can buy a glow fuel engine for much less," let me tell you that a 400-square inch electric powered model aircraft can easily be flown with either a \$30.00 ferrite motor or a \$100.00 cobalt motor.

Now comes the real problem. How do you size an electric motor? How are these motors rated for specific model aircraft applications?

With glow or diesel model engines, the

Belt drives, which provide the same mechanical advantage as a gear drive, have one specific advantage. The motor's rotation doesn't have to be reversed. The Modelair-Tech H-1000 belt drive can house one or two motors as shown here to fly models up to 1/4 scale.

PHOTO: BOB ABERLE

In this close-up of the Modelair-Tech H-1000, see the belt and gear assemblies. Note the elongated slots that permit easy belt tension adjustments. The unit can accept all types of motors.

PHOTO: BOB ABERLE

Amp-Air's new dual motor gear reduction drive is designed primarily for 540 and 550 type 05 motors, and the new SR Batteries' neodymium motors.

PHOTO: BOB ABERLE

cubic inch displacement of the engine is the reference point. In the infancy of model aircraft electric power, someone attempted to assign cubic inch displacement figures directly to identify electric motors, so that modelers would get an idea of how they related to more familiar glow engines in power output. It is common today to see electric motors "sized" as 035, 05, 15, 25, 40, 60 and 90. This type of designation holds to this day for the popular line of Astro Flight cobalt motors.

On the ferrite side, the basic can motor is known as a 540 or 550 type, which is close to an "05" in the cobalt series. The very popular German Graupner ferrite motors are classified as Speed 400, 500, 600, and 700, with various sub-categories of voltage ratings (providing a comparable cobalt range of 035 to roughly 25).

The new brushless DC motors, which are now available from three different sources, use both a wattage rating, as well as a cell count. For example, Aveox Inc. has five models available classified by the recommended number of battery cells (like 5-7 cells up to 9-16 cells).

In general, the electric motor ratings will eventually make the most sense if one uses the typical operating power (expressed in wattage) of the motor. The most familiar today are the output wattage ratings of the Astro Flight cobalt motors where the 035 is up to 90 watts (operating on five battery cells). The 05 operates up to 125 watts (seven cells); the 15 operates up to 200 watts (12 cells); the 25 is up to 300 watts (14 cells); and the 40 is up to 450 watts (18 cells). Beyond that is the 60 and 90 size motors for the really large models where 28 to 32 (and even more) battery cells are common.

This is only the beginning, because other variables come in to play with motor designations. For example,

Extremely adaptable, the Sonic-Tronics radial type motor mount can be adjusted to fit any motor up to a "15." It is designed so that the mount will break before the motor shaft is bent.

PHOTO: BOB ABERLE

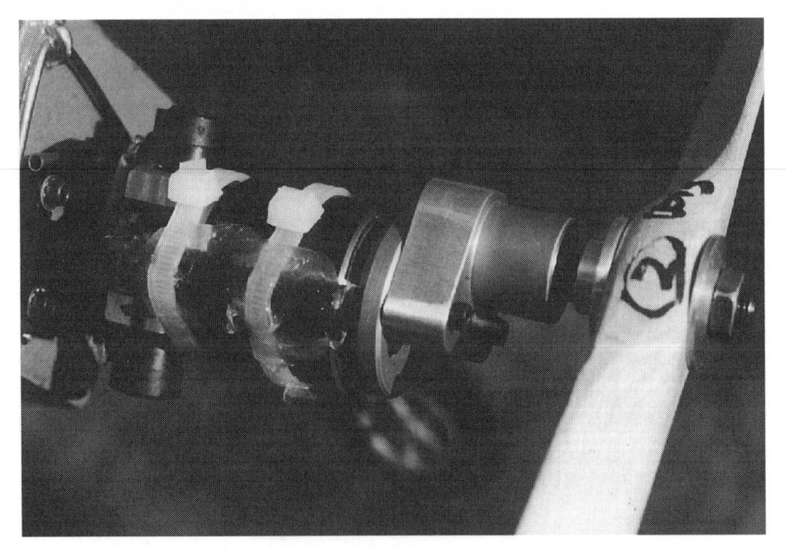

Here the SonicTronics mount holds an Astro Flight FAI geared 05 motor on the front end of the author's Old Timer Model. The nylon ties keep the motor in its mount, while the silicone sealer keeps the ties from moving.

PHOTO: BOB ABERLE

Unique among gear boxes is the Hobby Lobby 2:1 titanium unit. Its main advantage is that because it's a planetary gear box it is inline, with virtually no offset like most other gear boxes. It is used almost exclusively with the popular Graupner Speed 400 motors running on five to seven cells.

PHOTO: BOB ABERLE

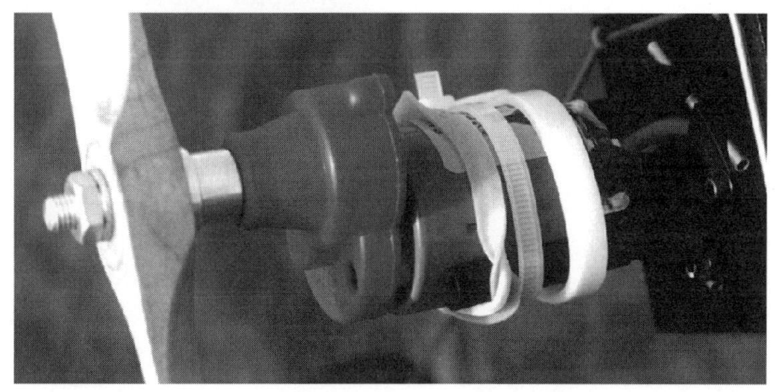

One of the most popular, and also one of the most inexpensive of gear boxes is the Master Airscrew unit that works well with a wide variety of small "05" motors. It's shown mounted on a Trinity Sapphire 17-turn ferrite motor.

PHOTO: BOB ABERLE

the number of turns in the motor windings and the gauge of the wire has a definite effect on the power (wattage). Astro Flight has one set of motors for sport use and another set for competition flying which is known as their "FAI Series" producing more power (and consuming more current!).

A final variable to motor selection is whether you turn the propeller directly or through a gear or belt reduction drive. In general, a direct drive system will produce high motor speeds, high current drain, and consequently short duration motor runs. Direct drive is popular with sprint type flying such as pylon racing and high performance electric powered sailplanes, where motor runs are often limited to as little as 30 seconds, pursuant to 8-minute total time flights (typical contest rules).

Running a propeller through a gear or belt reduction drive assembly can provide greater thrust (to lift heavier models) at more modest current drain, which allows for longer flying time on a single battery charge. General sport, scale and certain duration type model aircraft can benefit from reduction drive units.

Keep in mind that most add-on gear or belt drives cause a certain offset between the prop shaft and the motor shaft/casing. That can prove difficult at times since the motor ends up on one axis, while the prop shaft can be offset by an inch or more away from the motor. Providing fuselage or cowl clearance can be a problem. It is possible to obtain a planetary type gear drive for which there is essentially no offset. If you have a small fuselage, consider this type of gear drive.

When installing gear drives, you must consider the polarity of the motor (+ or –) which affects the direction of rotation. Adding a gear box will require that the motor polarity be reversed. With belt drives this is not necessary. There is also a possibility that the motor "timing" might have to be adjusted for optimum motor performance. A discussion of timing is beyond the scope of this book. However, you can consult the instruction manual supplied with your motor to see if such adjustments are recommended and how to do it.

It is advisable with most motors, and especially the ferrite types, to break them in before placing them in service. The best way to do this is to run the motor, without a prop, through several battery charges. You should periodically monitor the motor during this break-in. If the mo-

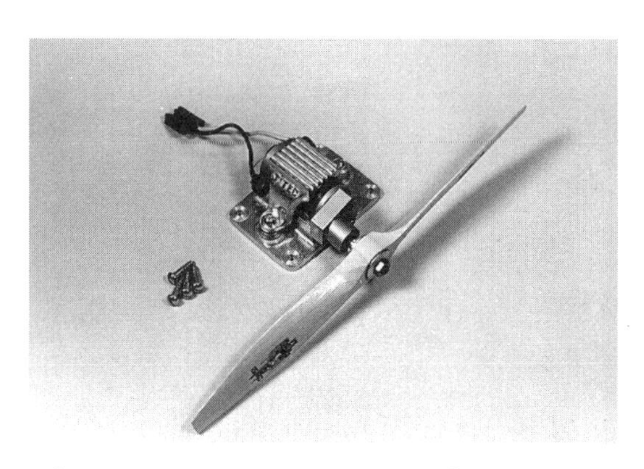

Electric motors can get very warm as they run. That's why this J-Tec motor test stand is made of heavy aluminum to act as a heat sink.

PHOTO: BOB ABERLE

Gear ratio determines the motor/prop combination. This Model Electronics Super Box unit is adjustable to provide 70 different ratios.

PHOTO: BOB ABERLE

tor case begins to get hot, stop the motor for a while to let it cool. The break-in period will generally allow the brushes to seat properly, which in turn will provide more power and efficiency.

Unfortunately, motors are usually sold without any form of mounting provisions. So, how do you mount an electric model in your airplane? Well, essentially all of the electric motors we use have a basic cylindrical shape. On most motors you will find tapped screw holes on the front plate which makes firewall type mounting quite easy.

In any form of motor mounting technique you must consider that electric motors get warm when running. The longer they run and the faster, the hotter they will get. It is, therefore, very important that all electric motors get some amount of air circulating around them to aid in the cooling process. If you firewall mount your motor, at least provide several front facing air intake holes and a passage for the air out beyond the rear of the motor.

A particularly neat, glass filled motor mount is offered by the SonicTronics Co. This mount is adjustable for various motor diameters. It also has a designed-in feature that will cause the mount to break away in the event of a hard landing. These mounts only cost a few dollars and it is a good idea to keep a few extras in your field kit. That "break away" feature can save many a bent motor shaft. Bending motor shafts, by the way, can prove one of the most costly and annoying repair problems in electric flight.

Some of the new and better designed gear and belt drive assemblies offer mounting flanges which can facilitate beam mounting similar to that employed on internal combustion engines. This puts the prop thrust directly on the mount, not on the electric motor. Other motor mounts can be fashioned from such things as nylon tubing which you wrap around the motor much like a clamp.

Electric motors can and often do generate impulse type electrical "noise" of sufficient magnitude that can actually feedback into the R/C system causing in-

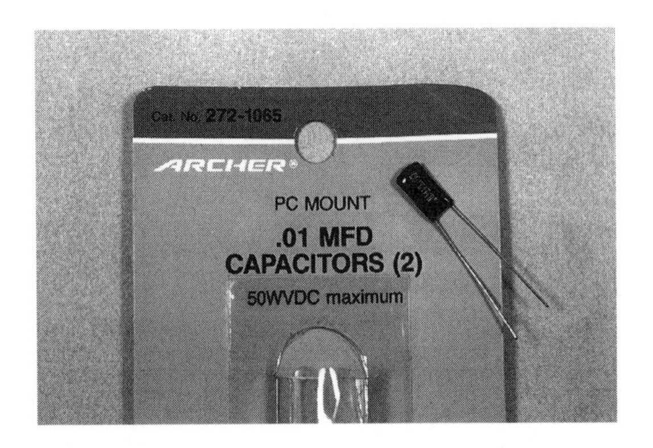

These Radio Shack capacitors are placed between the motor terminals and ground to help reduce generated motor noise which might interfere with the R/C system.

PHOTO: BOB ABERLE

terference to the controls. To suppress this motor generated noise it has been a common practice for years to place a small by-pass capacitor across the two motor terminals. In more recent times it is even more common to see three such capacitors used. One is placed from each motor terminal (positive and negative) to a ground connection (like the motor case itself). A third capacitor goes directly across both terminals.

Some motor manufacturers provide these capacitors already installed. On many of the ferrite "can" motors these interference suppression capacitors are installed inside the case (can) so the actual installation won't be obvious. In any event, if you don't see these capacitors, install at least one, or better still, all three as just described. An additional set of capacitors won't hurt. These capacitors can be purchased at a Radio Shack Store (P/N 272-1065, .01 MFD at 50 WVDC).

I'm sure that by now you are probably wondering how you might ever hope to select the proper motor, type of drive, prop, number of battery cells and capacity for any given model aircraft application. Until recently, the task was almost impossible and probably the single reason why so many modelers were reluctant to enter the world of electric powered flight.

Now, thanks to the age of computers and to some very sophisticated new software developments, it is possible to determine all of the necessary parameters to support electric powered flight, without the need for a lot of expensive and time consuming experimentation. Feed your model type into the computer program, tell it what you expect in the form of performance and you will receive the necessary recommendations that will enable you to purchase the best, motor, gear drive, prop, and battery (cell count/capacity) combination. I will get more into this in a later chapter.

props

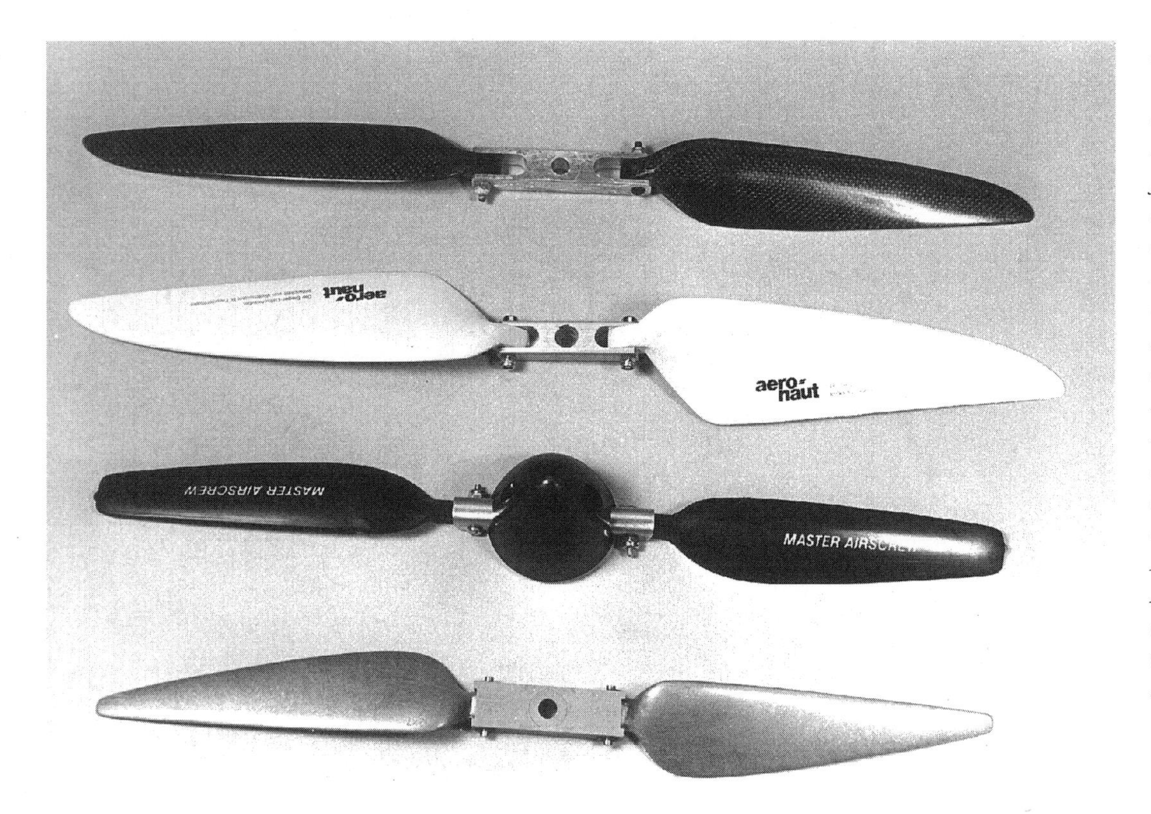

Electric power has made modelers more conscious of prop selection. Early on, folding props were quite popular, and continue to be for their ability to reduce drag when they fold back. They come in all sorts of shapes and material. From the top to the bottom: a Freudenthaler with its carbon fiber blades; an Aero-naut; a Master Airscrew which comes with a spinner; and a SonicTronics.

PHOTO: BOB ABERLE

Propellers, or simply props as they are more commonly called, are considerably different for electric powered flight operations. If you have flown glow, diesel or ignition type model aircraft engines in the past, the distinction from electric power props will become very obvious.

A word of caution is worth noting at the beginning of this chapter. Never attempt to use props intended for electric power on any other form of powerplant. The relatively smooth, vibration-free environment of electric power, in combination with generally much lower rpm, will permit props to be made thinner (with less weight) than would be the case for other forms of power. In addition, folding type prop blades are quite common with electric power. Again, this should be an application strictly for electrics. These blades can be set up to

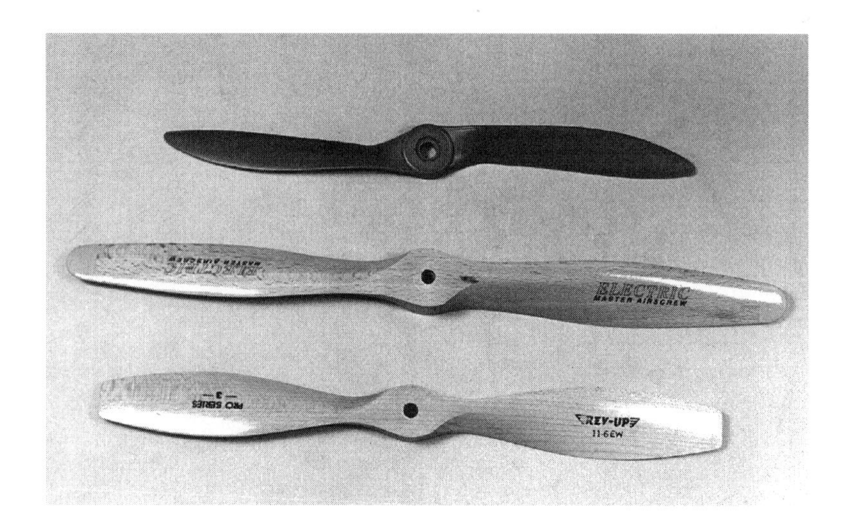

Standard "fixed" type, i.e. non-folding, props are now being designed specifically for electric motors. From top to bottom: an APC plastic; a Master Airscrew Electric wooden prop; and a Rev-Up wood prop (with extra wide blade).

PHOTO: BOB ABERLE

Graupner products are distributed in the U.S. by Hobby Lobby. These Scimitar folding props were designed for electrics and come complete with spinners. Individual blade sets for various diameters and pitches are available to replace the original blades for different thrust requirements, or to replace broken blades.

PHOTO: COURTESY HOBBY LOBBY

fold against the forward fuselage of the model aircraft every time the engine is shut down in flight.

In addition to the obvious advantage of streamlining the model in flight and improving overall performance, there is another advantage to consider. Motor shafts attached to the armatures of electric motors are usually not made of hardened steel material. Many electric powered models, especially the sailplane types, are usually hand launched and, as a result, don't have or need landing gears (and wheels). Any kind of a hard landing can easily cause a bent motor shaft which can prove to be an expensive and time consuming repair item, especially when parts have to be ordered. The use of a folding prop usually helps prevent bent motor shafts, in most cases.

Another benefit of folding props is that the blades are less apt to be broken on rough landings. Folding props may initially cost more, but can save you a lot of money in the long run.

Most folding props are supplied unassembled. The assembly process takes little time but should be done with care. Always make sure the two blades are oriented on the hub properly. Also make sure that the blade mounting screws are tightened per instructions. If lock type nuts are called for, make sure you use them. If "E" type retaining washers are used, make sure they are positively snapped in place in the prescribed slot.

Most electric motors will require a prop adapter as shown here. It fits over the motor, or gear box shaft. Then the prop is bolted to the adapter.

PHOTO: BOB ABERLE

Several of the brands of folding props offer combinations of blades and hubs which are sold separately. You can obtain very close to the diameter and pitch that you need for your specific application by following a provided conversion chart. One of the popular manufacturers of this type of multi-purpose folding props is SonicTronics Inc. They make a good product, but their hubs are not marked with a part number. So when you buy a new hub from SonicTronics make sure the first thing you do is write the assembly number on the hub (as listed on the packaging) with something like a Magic Marker.

Several other things should be kept in mind when using folding props on electric motors. It is generally a good idea to use a speed controller that includes a prop brake feature. The brake acts electrically to stop the prop rotation. This is especially important when using a direct drive motor. Without a brake, it is sometimes difficult to stop the prop from turning after the motor is shut down. If it keeps turning, it can't fold. When using gear or belt drives, brakes are not considered advisable.

Another thing to look for is a speed controller with a slow start feature. On the larger diameter folding props there is a remote chance that the prop blade could strike the wing leading edge at the instant it starts rotating. Having a slightly slow start allows the prop blades to first extend before they have a chance to reach any real speed.

Not all props used for electric power are folding types. Wood and plastic (one piece) props are also popular. In general, the prop pitch is usually a high number, i.e., a 12-inch diameter by 8 pitch (referred to as a 12–8) is quite common. You will also see 15–10 props used for many electric flight applications. Gear and belt drive reduction units prompt the use of higher pitch props. Many of the higher pitch props which were originally designed for gas engine service are

This SonicTronics' folding prop has rubber bands installed to defeat the folding ability of the blades in flight. This is a specific requirement for props in the Electric Flight category of Old Timer events.

PHOTO: BOB ABERLE

much too heavy for electric power use. As a result, we are now seeing electric power specialty props showing up on the market.

The Windsor Propeller Co. now offers a line of Master Airscrew electric wood props that have the popular diameters and pitches and also the proper camber needed for electric motors. These props have been thinned considerably to save weight.

When purchasing various electric motors don't be surprised if they don't come with a prop adapter assembly. Not every motor manufacturer is that thoughtful. Adapters are available for most motors and can be installed on the output shafts in a matter of a few minutes.

batteries & chargers

Figure 6-1

Typical nickel cadmium (Ni-Cd) battery pack cells are always connected in series for electric flight operation.

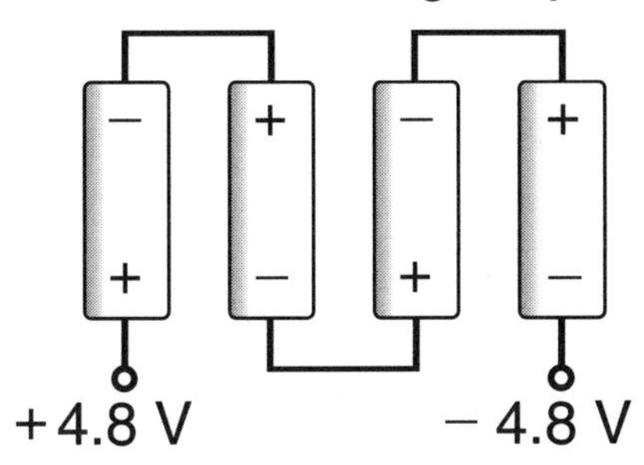

Each battery cell is 1.2 volts nominal. In series connections, voltage is the sum of each cell's nominal voltage.

Batteries are the life blood of electric powered flight. To a degree, the battery is very much like the fuel you use to power reciprocating engines. Beyond that simple analogy, the differences are considerable. If you want to add flight time (duration) to an electric powered model aircraft you must go to higher capacity battery cells. That can be done easily, but at a certain weight penalty. The first consideration is that when you add capacity, you add weight. Obviously there is a limit to how much capacity you add before you reach the point where it would be impossible for the model to leave the ground.

Electric powered models can be made to go faster in certain cases by selecting a different size prop or by changing the type of drive (direct or gear/belt). The primary way to increase speed is by increasing the voltage going to the electric motor. Increasing the voltage is accomplished by adding more battery cells, in series, which again will cause more

Battery cells are never connected in parallel, i.e. all positive terminals connected to each other and the negative to each other. battery capacity, i.e. duration, must be gained by the individaul cell capacity rating.

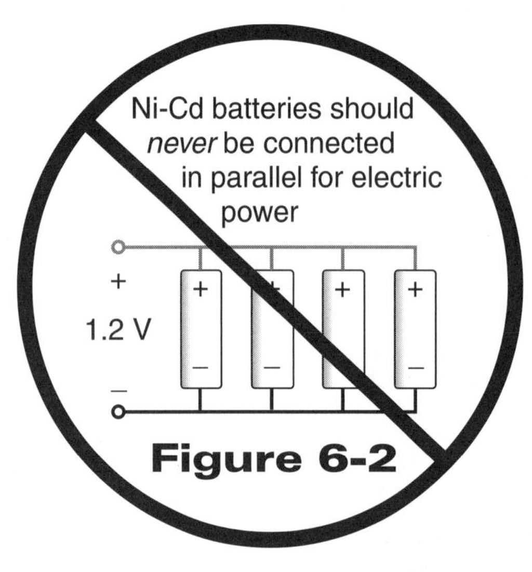

weight.

When you increase the voltage you also increase the current being drawn from the battery and hence, the operating time (duration) is reduced. In both cases the situation is one of increased weight as a trade-off to increased performance (longer duration and/or faster speed). Everything you do in electric power is always a compromise between expected performance and the total flying weight that your aircraft can tolerate.

Since the mid-sixties the most commonly used battery for electric powered flight has been the nickel-cadmium variety which most of us refer to as a "Ni-Cd" (pronounced "ni-cad"). The very first operating rule for Ni-Cd battery cells is to connect them only in series, never in parallel. By series I mean a continuous link where the positive and negative terminals are connected end on end (see **Figure 6-1**). The nominal voltage of a single Ni-Cd battery cell is always 1.2 volts. Therefore, when four of these cells are connected in series, the total battery pack voltage will be 4.8 volts (or 4 times 1.2 volts).

In regular electric powered flight service, the battery cells are always operated in a series loop. While flying, the battery cells are being discharged. After the flight is concluded, these same cells must be recharged (have the energy put back in) and again this is done always with the cells in a series loop.

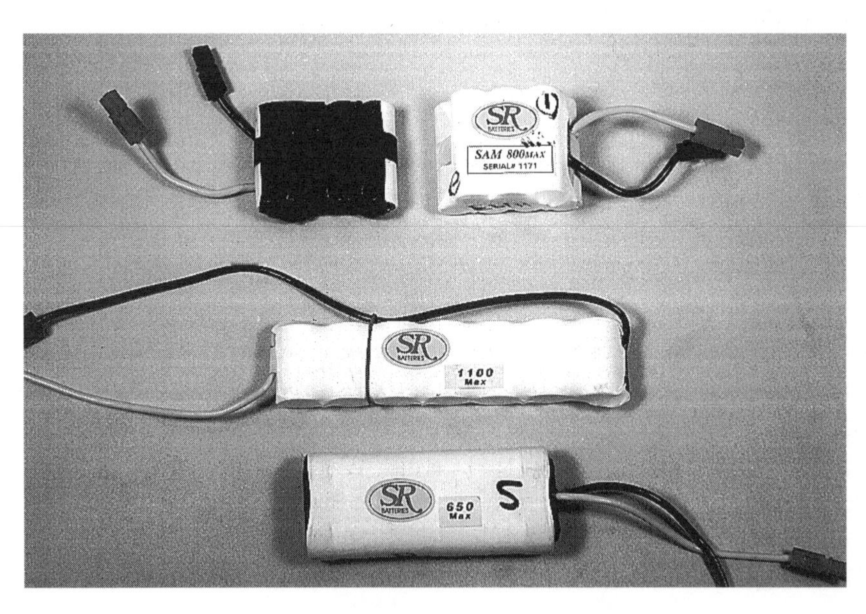

These are different types and sizes of Ni-Cd battery packs. At the top left is a 4-cell 1250 mAh pack. Top right shows a 7-cell, 800 mAh pack. At center is a 7-cell 1100 mAh pack, and at the bottom is an 8-cell 650 mAh pack.

PHOTO: BOB ABERLE

To increase the capacity of Ni-Cd battery packs you simply buy new cells which are rated at a higher capacity. Never place the cells in a parallel connection to gain capacity (refer to

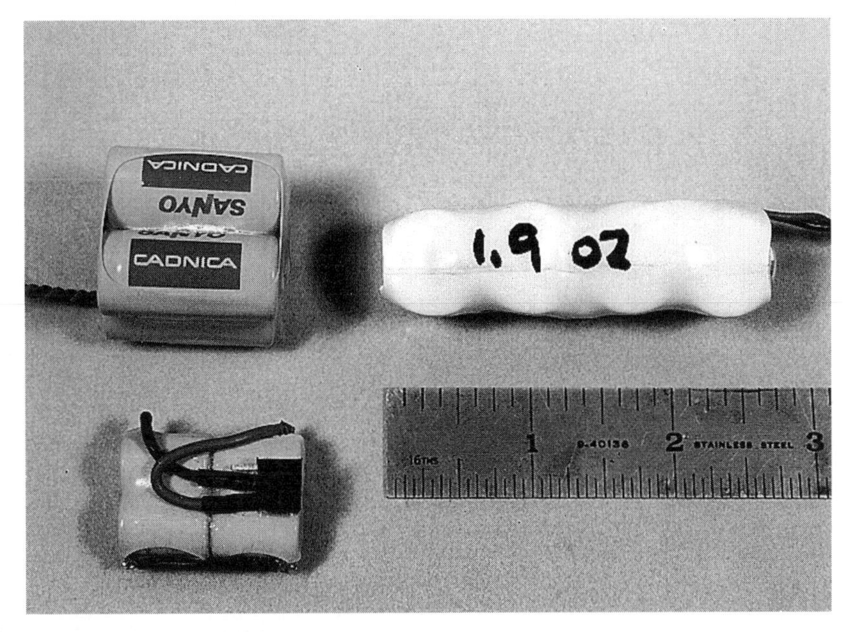

There are also small battery packs for electric freeflight and indoor electric R/C models. At the top left is a Sanyo 4-cell 270 mAh pack; at the right is a 4-cell 225 mAh pack. The bottom is a tiny 4-cell 50 mAh pack.

PHOTO: BOB ABERLE

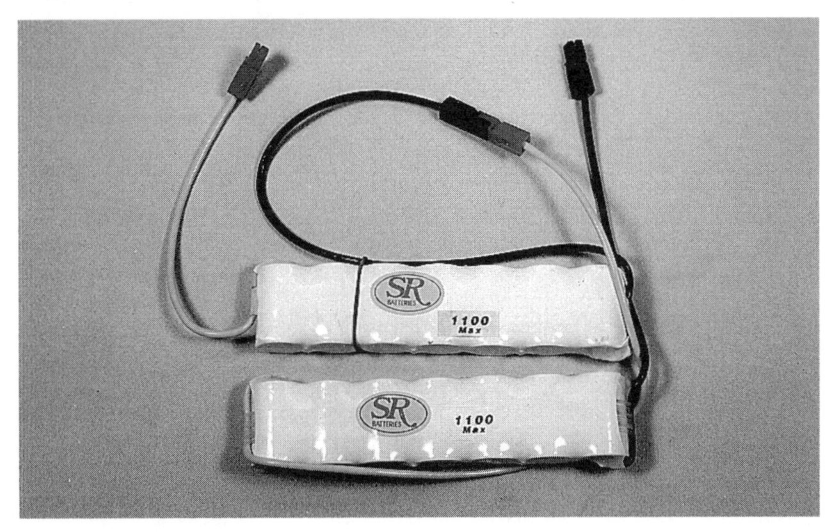

In this case, two individual SR 7-cell, 1100 mAh, 8.4 volt battery packs have been hooked together in series to provide a 14-cell 16.8 volt battery pack. The capacity is still 1100 mAh. All 14 cells can be charged at the same time.

PHOTO: BOB ABERLE

Figure 6-2). Ni-Cd battery cells are commonly rated in milliampere hours or ampere-hours. Milliampere hours (or mAh) is the most common terminology. As a point of information, 1,000 milliamperes is equal to one ampere. Taking that a little further, a Ni-Cd battery cell rated at 1100 milliampere hours is the same as saying a cell rated at 1.1 ampere hours.

What does this capacity rating actually mean? Let us use that 1100 mAh battery cell as an example. First, fully charge the cell (I will explain full charging later in this chapter). Next apply a load to this cell equal to the rated capacity. In this case the rating is 1100 mAh, so the applied load would be 1100 milliamperes (abbreviated as mA). Once the load is applied, start a timer going and determine how long it takes for the cell voltage to drop to 1.0 volts. If the cell was accurately rated at 1100 mAh, it should take exactly one hour to reach 1.0 volts.

The same situation would be true if you had four Ni-Cd battery cells connected in series for a nominal 4.8 volts. A load of 1100 mA would be applied and the timer started. Again if the rated capacity was accurate, it would take one hour for the battery voltage (sum of four cells) to reach 4.0 volts (1.0 volt per cell

The ACE DMVC (Dual Metered Vari-Charger) is an example of an adjustable low output battery charger. It has two adjustable outputs, each from 0-250 milliamps (mA). PHOTO: BOB ABERLE

times four cells in series).

There are many battery testing devices on the hobby market today which will allow you to either directly or indirectly establish or verify the capacity of your Ni-Cd battery pack. The type of service you give your battery will have a lot to do with how long it will last. Ideally you want your pack to handle several hundred recharges over the course of several years of flying. That type of battery performance is quite possible. On the other hand, there will occasionally be battery packs that will have one or two cells die in just a few short months of service. Keeping logs on your battery packs and recording capacity figures every few months is an important part of routine battery maintenance.

What type of Ni-Cd battery cells should you look for? The popular brands today are Panasonic, Sanyo, and SR. The modeler, therefore, has several choices when it comes to battery packs. First you can buy individual cells and make up your own battery packs. A second choice is to purchase ready-made packs from SR Batteries Inc. This is the only company that selects and matches cells prior to assembling them into packs. They will make the pack in any configuration you desire with the wire gauge, length and connectors to your specifications.

Basically Ni-Cd battery cells are offered in three different types. Using the Sanyo brand as an example, the standard Ni-Cd cell, which is usually in a yellow jacketed sleeve, is intended for low power consumption levels, primarily for the operation of R/C systems (receivers and transmitters). They can, however, be used to power some electric flight systems if the motor current can be held to reasonably low levels (under 10 amps). Small electric powered R/C and freeflight models, operating both outdoors and indoors, can acceptably use these standard Ni-Cd cells. Although they are not intended for fast charging, they can handle it within certain limits.

> **Overnight Charge Rate**
> **cell capacity ÷ 10 = charge current**
> **e.g.**
> **1100 mAh cell ÷ 10 = 110 mAh current**

The most popular Ni-Cd cell is the low

internal impedance type intended for general electric powered flight. It is the type similar to the Sanyo SCR cells, which generally have a red protective jacket. These cells are available in such popular sizes as 600, 800, 1000, and 1400 mAh capacity ratings. More recently a very heavy duty type, the SCRC cell, has appeared with a rating of 1700 mAh. Any of these cells are capable of supplying very high current levels, upwards of 50 amps in some cases. Probably the better average current drain level should be held to between 25 and 30 amps.

Not only do these battery cells handle fast discharging (in flight), but they also permit very fast charging to restore the power. Fast recharging means that you can get back into the air sooner. In the SR Batteries line, their low internal impedance cells can be found in their "MAX Packs."

The final type of cell is characterized as having high internal impedance, and is similar to the Sanyo SCE or simply "E" type. They generally have a yellow casing with black lettering. These cells are known to have a higher capacity for a given weight. Unfortunately, the problem when using this type cell is that you are generally limited to motor currents of approximately 15 amps (20 amps at best!) which eliminates many applications right from the start. These same "E" cells also prefer lower charge rates than the SCR type. But again, for certain applications the higher capacity can be worth it. In the SR Batteries line, these high internal impedance cells go into their "Magnum" packs.

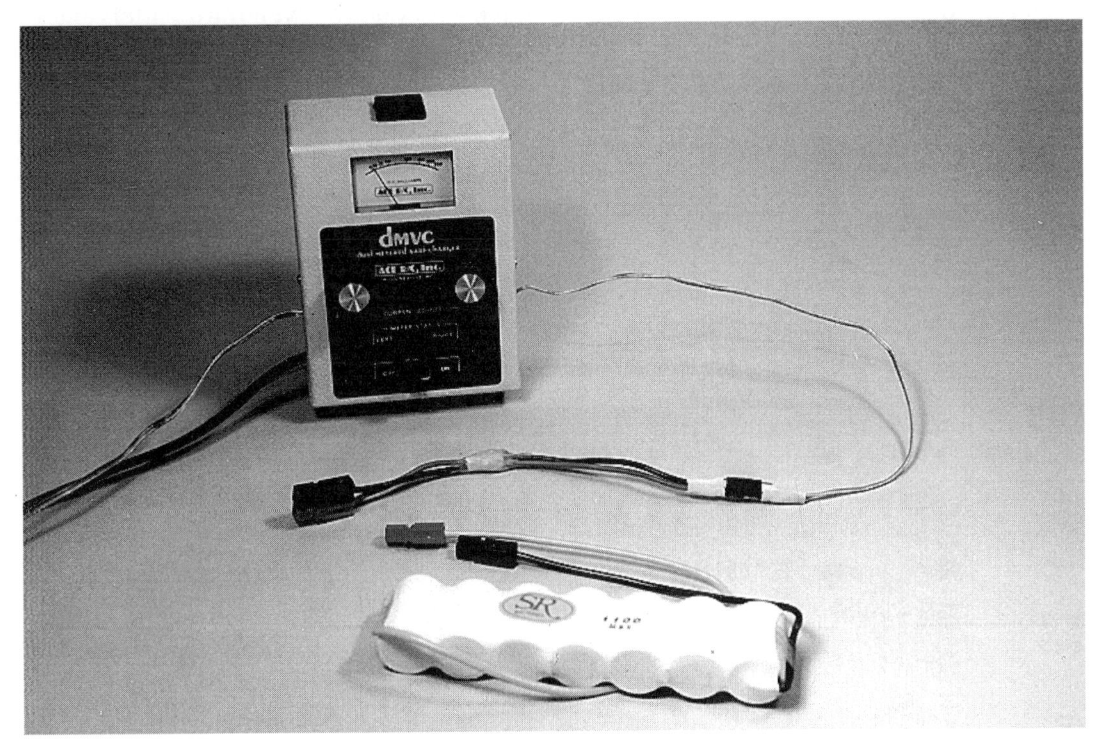

Once a month it's a good idea to charge over night at the C/10 rate. This Ace DMVC (Dual Metered Vari-Charger) can be adjusted to charge this 7-cell 1100 mAh battery packat 110 mA which would be the overnight charge rate.

PHOTO: BOB ABERLE

In the previous chapter, electric motor heat was mentioned as a concern. Battery heating is likewise a concern which must be addressed by electric power enthusiasts. During the flight, battery packs will tend to get hot, with the degree they heat up being directly related to the current drain of the electric motor. A motor drawing 20 amps current over a 5-minute period will get the battery warm, while another motor drawing 35 amps current for 2 minutes will get the battery quite hot. Still another battery supplying 45 amps to a motor for just 30 seconds will be almost too hot to handle. After landing your aircraft, it is advisable to wait until the battery pack has sufficiently cooled down before attempting to recharge it. Recharging a very hot battery will greatly reduce its life span.

Good model engineering practices dictate that you provide some form of airflow over the battery pack. This means providing an air scoop at the front of the battery pack that directs incoming air around the battery. Equally important is an air venting hole behind the battery pack to allow the hot air to exit the aircraft. Even after making such air cooling provisions you may still find your battery pack quite hot after landing your model. To speed up the cooling process you could blow air supplied by a small 12-volt fan over the battery. Radio Shack has several small fans listed in its catalog at reasonable prices.

Another suggestion is to physically remove the battery pack after landing and charge it outside the aircraft. I prefer this approach and as a result have never resorted to using a charging jack that would permit charging of the pack while it remained inside the model. Should you have a second, or even third battery pack, it becomes a simple matter to have one cooling down, while another is being recharged and a third is ready for a flight. Having extra battery packs also spreads out the duty cycle, allowing the packs to cool down properly and also provide more flying time.

Another question relates to how to treat batteries when not flying, or when they are stored. There are two choices: leave them discharged until ready to go flying the next time, or recharge them with an occasional "topping off" if there was a long lapse between flying sessions (in other words, store them at close to full charge!).

My personal approach for years has been to recharge my battery packs when returning from the flying field, and do it

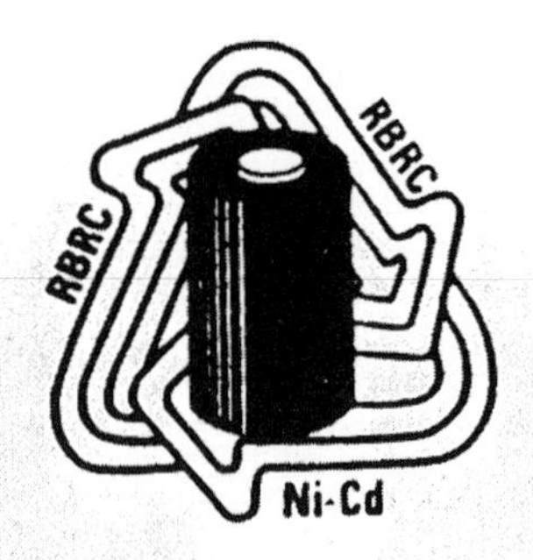

RECHARGEABLE
BATTERY
RECYCLING
CORPORATION

at the overnight charge rate. What is the overnight charge rate? It is the rated capacity of the battery divided by 10 and expressed in milliamperes (mA). For example, if I have a battery rated at 1100 mAh I would divide the 1100 by 10 which equals 110. That pack would get a charge current set at 110 mA for an overnight period of at least 12 hours (up to 24 hours wouldn't hurt!). By using that slow rate, I'm actually conditioning my batteries so that they will accept the fast charge rate better the next time I go to the flying field.

Adjustable low output battery chargers have been sold by ACE R/C Inc. for many years. If I'm unable to fly during the winter due to bad weather, I generally will recharge at the C/10 rate about once a month until flying resumes in the spring.

Something else to consider when talking about batteries intended for electric powered flight is the effect of conditioning prior to the first flight of the day. Regardless of the normal use of your bat-

DAD Inc. is one of many companies that offers a peak detection charger. It can handle batteries from 4-10 cells with a charge current adjustable to 5 amps. It will auto-revert to trickle, and can operate on both 12 volts DC or on 110 VAC.

PHOTO: BOB ABERLE

SR Batteries Smart Charger/Cycler, or SC/C, is an extremely sophisticated and capable charging device. It can peak detect charge, slow charge, test batteries plus a variety of other applications.

PHOTO: BOB ABERLE

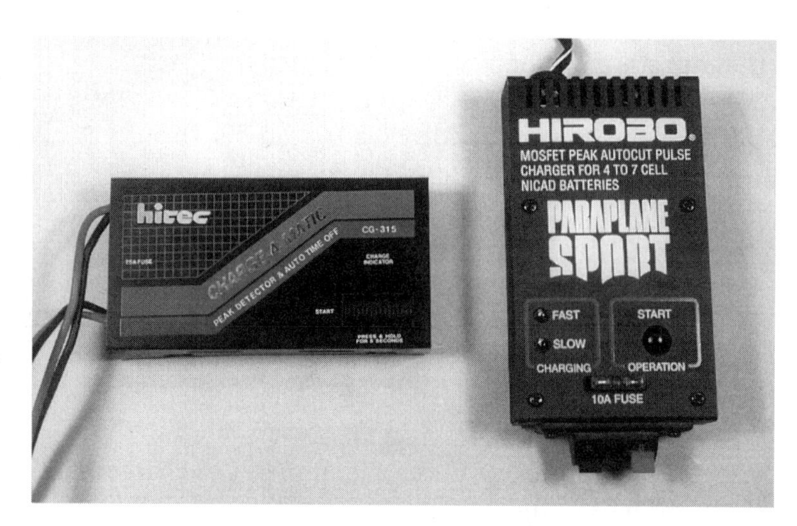

The Hitec Peak Detect Charger, at the left, is a simple and inexpensive charger that handles 5–7 cells at 3 amps. The Hirobo Peak Detect Charger, at the right, handles 4–7 cells, and is sold with the Hirobo/Altech Marketing Paraplane Sport.

PHOTO: BOB ABERLE

teries the very first flight of the day will never be as good as subsequent flights on that same day. It usually takes one discharge and recharge cycle to condition the batteries to their full potential. If you are just sport flying, there is nothing wrong with a "soft" first flight of the day. However, if you are flying in competition and are not allowed any test flights, it is advisable to top off your charge when arriving at the field, run the motor once taking the battery pack all the way down and then recharge it. Your first official flight in a contest will then be the very best you can do with your equipment.

A lot will be said in various articles and electric columns about making your own battery packs from loose cells. "Rolling your own" is a definite possibility. I've done it with good results, but it does take time and you do have to develop certain skills in assembly and soldering to do the job correctly.

My personal preference is to go to a custom battery supplier, like SR Batteries Inc. where I can specify the pack con-

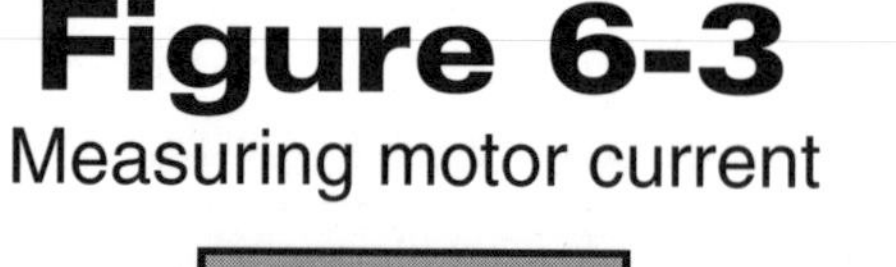

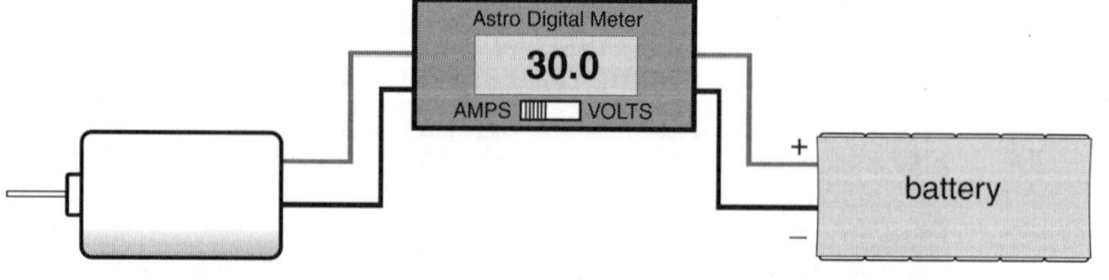

To understand efficiency, it is necessary to measure motor current to get the best match between motor, prop, and battery. The diagram shows a way of determining that using the simple Astro Flight Digital Meter.

PHOTO: BOB ABERLE

figuration, cell type, wire size and length, and type of connector. If after its first use I feel the pack isn't up to my standards I have someone to complain to besides myself. If I buy loose cells, assemble them myself and obtain poor results, I'm stuck with the end product. Having someone do the battery assembly work also allows me more time for building and flying.

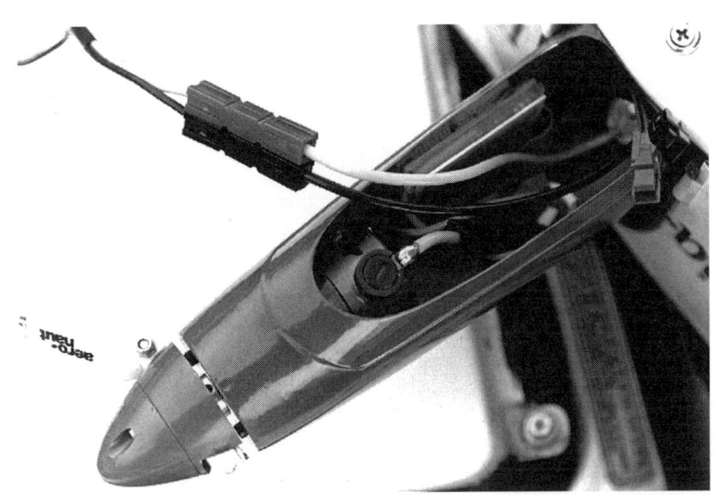

The canopy, in this case, has been removed to gain access to the battery connector for a typical charging routine. With most chargers, it will take about 20 minutes to recharge.

PHOTO: BOB ABERLE

Many people have asked for a simple formula that they might use to determine the theoretical motor run time on a battery pack of stated capacity, when only the motor current is known. What I generally use is the battery capacity expressed in amp/hours × 60 and then divide that by the motor current in amps.

For example I have an Old Timer model whose motor current is 30 amps. **Figure 6-3** indicates how you can measure your motor current. I'm using 800 mAh batteries, so the amp/hour rating would be 0.8 amp times 60 which equals 48. Divide that by 30 amps and you get 1.6 minutes (or a little over 90 seconds). Remember, that is just a rough calculation because, in reality, the prop will tend to unload somewhat in flight causing the motor current to go down. The actual battery capacity may be different from that stamped on the cell casing. All of these factors introduce variables to the calculation, but still in all the formula works as a good guide to motor run duration.

Before we leave the subject of batteries, I wanted to mention another type of battery cell that is beginning to show up in hobby advertisements. They are called, Nickel-Metal Hydride (NiMH). The claim is that you will be able to obtain more flying time without added weight. That claim is true, but the fact is that this type of battery cell cannot handle the high current loads demanded by electric motors, nor can they be recharged at a high rate. They are primarily intended to power R/C systems, both the transmitter and receiver batteries and to be charged at the more usual overnight rate.

The very last item on batteries: how do you dispose of them safely? If you are "ecologically" concerned, and I hope you are, you will be interested to know that after January 1, 1995 battery manufacturers who are members of the Rechargeable Battery Recycling Corp. (RBRC) will affix a special symbol to all new battery cells. RBRC has set up battery recycling centers throughout the country for your safety and convenience. They will gladly accept and properly dispose

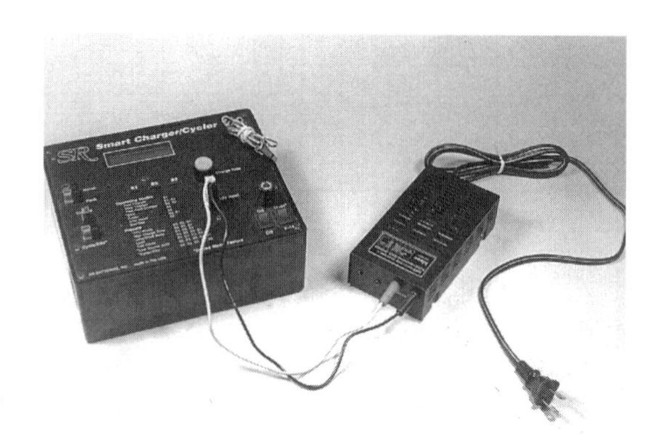

More sophisticated chargers can use a dual power input source, 12 VDC or 110 VAC, like this SR Batteries SC/C shown here being operated from a 110 VAC source.

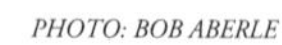
PHOTO: BOB ABERLE

of your battery cells. Most local recycling centers will also accept batteries, nickel-cadmium as well as alkaline, for disposal.

How do you charge batteries? To fully enjoy electric powered flight you must have a charger that works at a fast rate and preferably one that is automatic. An automatic charger is one that can be left unattended while it fully charges the battery pack and then turns itself off (or reverts to a so-called trickle level). You want to be able to charge one set of batteries while flying with another set without having to "baby sit" your battery charger. The most popular way to charge Ni-Cd battery packs today is by the technique known as "peak detection."

As Ni-Cd battery cells are being charged, the voltage being applied to cells continuously increases. Eventually, a point will be reached when the voltage, under charge, peaks or, in other words, doesn't go any higher. Shortly thereafter, the voltage will actually begin to decline. At that point of peak voltage, the Ni-Cd battery cell is defined as being fully charged. You could charge a Ni-Cd battery while monitoring the charge voltage with a reasonably accurate digital voltmeter. It will be obvious when the peak is reached after which the voltage begins to drop off. At that same point the battery will begin to get warm, since it is beginning to go into the overcharged region. If you continued to charge, the battery will eventually be destroyed by excessive heat build-up.

Fortunately for us, our electronic engineering friends have developed circuits that sense the peak voltage under charge. After the peak has been reached and the voltage drops off only a few tenths of a volt, the charger is shut down automatically. Hence the expression, "peak detection charger." Some chargers, after cutting off the high rate of charge, will revert to a low, or trickle charge rate, which can help to keep the battery "topped off" prior to the next flight.

Most chargers are offered today with the peak detect or automatic shut down feature. A good peak detect charger can be purchased for about $100 to $125. The very best state-of-the-art charger could run up to $400.

One of the ground rules to keep in mind when charging your batteries is to know the importance of establishing the acceptable charge current for any given type of Ni-Cd cell. It is the one decision you must make before turning on your battery charger. The choice of charge current involves using the rated capacity of the battery to be charged. If you have a pack made up of 1100 mAh rated cells, that is the same as 1.1 amp/hr. Typical fast charge capable SCR cells like three times their rated capacity for the charge current. That would work out to 1.1 times

3, or a charge current of 3.3 amps. If your battery was fully depleted of charge, it would take approximately 20 minutes at that 3.3 amp. charge current to reach full charge. If using a peak detect charger it would likely trip out around that 20 minute point.

If you are using the high capacity or "E" type cells, they generally like a lower charge current. The rule of thumb for fast charging "E" cells is to use two times the rated capacity. If you were going to charge "E" type cells rated at 1000 mAh or 1.0 amp/hour, the charge current would be set at 2.0 amps (two times 1.0 amp) in which case complete charging would take more like a 30-minute period. Generally speaking, SCR fast charge cells are more tolerant of high current drains, have slightly lower capacity, but can be recharged quickly. On the other side, "E" type cells can't tolerate a high current drain, but do have a higher capacity, yet unfortunately take longer to recharge. As you can see there are always compromises to be made in electric powered flight for particular applications.

The standard type Ni-Cd batteries shouldn't be fast charged, but most modelers will still take their chances, especially when these cells are used for the very small electric powered aircraft. I have had luck using the "two times capacity" charge rate on these cells, although I suspect the manufacturers would prefer you didn't go higher than "one time" the rated capacity.

Any of these three types of Ni-Cd batteries can be charged periodically at the overnight charge rate which is the capacity divided by 10. A 1000 mAh battery would be charged overnight at 100 mA (1000 divided by 10). Charging at this rate helps to equalize all the cells in a pack, allowing them to recharge at the fast rate more evenly.

As far as recommended peak detect battery chargers are concerned, Astro Flight Inc. offers a line of five different units. They range in price from $65 to $200. The least expensive unit can handle 6- or 7-cell packs at up to 4.5 amps charge current operating from AC or DC (12-volt) power sources. The most expensive unit in their line (Model 112 PK) can charge packs up to 36 cells at 5 amps, but only from a DC power source. Their $139 unit (Model 110XL) typically charges up to 16 cells. All of the current Astro Flight chargers offer the peak detect feature, the charge current is adjustable and is monitored on an integral ammeter.

As this book was being written, it was learned that sometime in early 1995 Astro Flight would be introducing a complete new line of battery chargers. Please keep that in mind when comparing the references in this book with any new

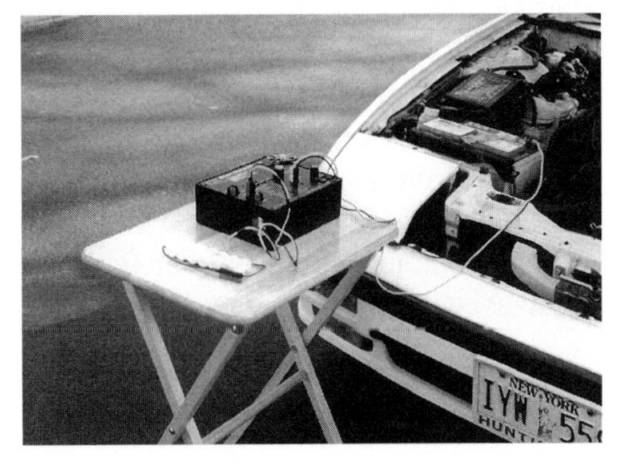

The SC/C charger uses a car battery as a 12 VDC power source. It has a sensor that monitors the car battery, so you can maintain enough power to start your car.

PHOTO: BOB ABERLE

brochures or catalogs.

Probably the very finest all around Ni-Cd battery charger currently on the market is the SR Batteries Inc., "Smart Charger Cycler" (or simply SC/C). This is a state-of-the-art micro computer controlled device which can perform just about any conceivable Ni-Cd battery charge/discharge function. It normally operates from a 12-volt DC power source, but an external AC power supply can be added as an accessory item. Prices for the SC/C start at $295.95 with the AC power supply as an optional accessory. I have used one of the first prototypes since early in 1992 with great success. Since it is micro computer controlled, it is possible to make periodic software upgrades to this system to keep adding valuable features, which makes your initial investment well protected over time.

Time would not permit a full technical description of the SR Batteries SC/C. I did review this unit in two separate articles which appeared in *Flying Models* magazine. Back issues may still be obtained by writing to Carstens Publications. The specific issues you may want to read are July 1992 and the software upgrade follow-up which appeared in the April 1994 issue.

Very basically, the SC/C provides an LCD screen with digital readout capabilities. When you leave this charger unattended, it will tell you a story on your return. For example, it will tell you the peak voltage attained under charge and how long it took the battery to get to that point. That, by the way, is a very important battery parameter worth keeping track of. As batteries tend to deteriorate with age, the peak charge voltage tends to increase. If you keep a record of the rise in the peak charge voltage, it will help determine when it is time to retire the battery pack.

By using tiny thermistors (temperature sensing devices) which you insert into your battery packs, the SC/C is able to control all the "thermal" requirements of your battery charging. For example, if you attempt to recharge a battery that the SC/C feels is still too hot, a message comes up on the LCD screen, "Pack Too Hot!." The charger will not allow itself to be turned on at that point. As the pack cools off, a point will be reached where the SC/C finally turns itself on allowing the pack to be recharged because the temperature has reached an acceptable level.

The SC/C also has a built-in safety feature that prevents you from recharging if it feels that the source voltage (your car battery) is too low. This way you can't fly all day and then find out there isn't enough battery power remaining to start your car.

The SC/C also has the capability for discharge testing battery packs at loads selectable up to 1 amp. It has an Expanded Scale Voltmeter (ESV) for checking R/C batteries at the field. I would suggest you either obtain copies of the referenced back issues of *Flying Models* or write to SR Batteries Inc.

controllers

Now that you have read about the motors, props, and batteries, the final ingredient to electric powered flight is how to control the motor system. For the most part our discussions on electric flight assume the use of a radio control system. I do expect to mention briefly electric powered freeflight (non-R/C) applications later in this book. For now the discussion will center on how to use your radio control system to control the electric motor in flight.

The types of motor controls possible include simple switches, relay switches, and solid state speed controls. Switches provide basically an on/off control. Speed controls provide a range of motor speeds from completely off, all the way up to full power, much like the variable speed carburetor works on an internal combustion engine. Modern electronic technology has essentially replaced the switches and relays with solid state, reliable, lightweight and relatively inexpensive speed control (motor throttling) devices.

In **Chapter 3** I briefly mentioned the use of relay type switch controls. It is only fair to elaborate further on these devices because they can offer the electric flyer a very inexpensive means of controlling the motor. There are three different brands of relay switches presently on the market that I am aware of. Each of the three contains the comparable electronic circuitry of an R/C servo, which means these relay switches take the place of a separate throttle servo. In each case, a servo type cable/connector (compatible with your particular R/C system) exits the relay switch and plugs directly into the throttle port of your R/C receiver.

One of the power leads going from the battery to the motor is passed through the relay contacts. Operating the throttle control stick on the R/C transmitter (from the low to high position) will close the relay contacts, thereby turning the motor on. To turn off the motor, reverse the procedure. Remember, you obtain on or off, but no speeds in between.

The first of the three relay switches is supplied by High Sky R/C Accessories Inc. It is relatively small in size and weighs only 1.1 ounces. The relay contacts are rated at 20 amps, but I have used

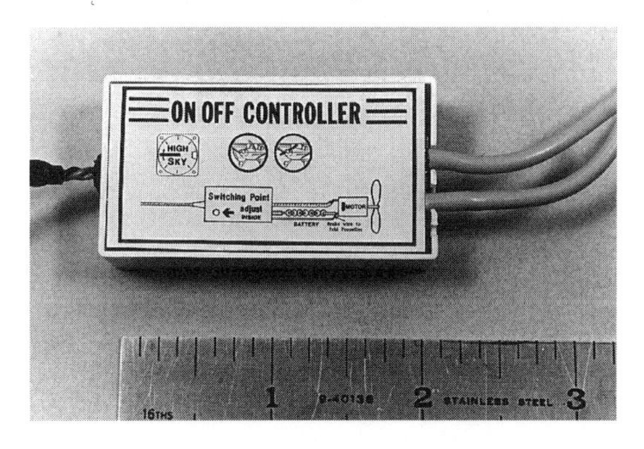

This High Sky On/Off Controller is an on/off motor relay switch that plugs into the receiver's throttle channel. It allows the throttle stick to turn the motor on or off.
PHOTO: BOB ABERLE

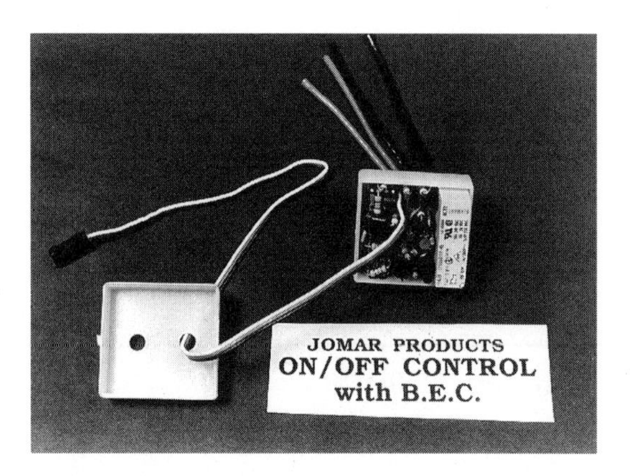

The Jomar On/Off Controller is a relay control which also includes a B.E.C. It can handle up to 25 amps current.

PHOTO: BOB ABERLE

this device at up to 25 amps for long periods of time without doing any damage. One thing must be kept in mind: while the relay is energized, approximately 50 mA of current will be drained from your R/C system airborne battery pack.

A second, and very popular relay switch is offered by Jomar Products Inc., which was recently acquired by Electronic Model Systems. This device is also rated at 25 amps (or 200 watts total power) and is small in size and light in weight. The added feature of the Jomar unit is a battery eliminator circuit (or "BEC"). If your motor battery has between six and eight cells, that same battery can be used to power your R/C system. In other words, the motor battery powers both the electric motor and the R/C system.

What happens in practice is that a regulator circuit supplies the nominal 4.5 to 5.0 volts to the R/C system and when a pre-set minimum voltage is reached, the relay opens, turning off the motor, but still leaving considerable power in the battery to operate the radio system. This will permit a safe landing.

A third relay switch is offered by the Robbe Company (now distributed and sold in the U.S. by Pica Models) and is designated as their Model No. RSC 200. Again it is a small device with a weight of just one ounce. Although this relay switch does not have the BEC feature, it is capable of handling power sources up to 25 volts and 50 amps, which is a remarkable rating for such a small device. Keep in mind that all three relay switches provide only on/off control.

The most popular means of electric motor control today is the solid state speed control. These controllers can be categorized into three types: 1) frame rate or low rate; 2) high rate; and 3) high rate with micro computer control. The latter will probably end up being the most popular as more enter the market place. However, there are numerous controllers of all three types presently on the market; therefore, it is worth discussing all of them to give you a basic familiarity. Actually, a fourth type of controller exists which is dedicated strictly to brushless DC motor operation. I'll discuss that briefly at the end of this chapter.

Frame rate controllers were probably the first to come along and have been on the market for quite a few years. Some

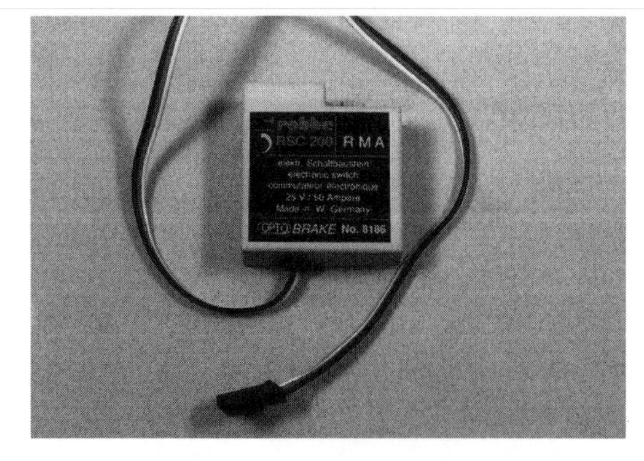

The Robbe RSC-200 is an on/off controller, but it can handle up to 50 amps. It weighs just 1.0 ounce.

PHOTO: BOB ABERLE

Small size electric power models demand small components, and the #217 Astro Flight Frame rate controller will comfortably fit in small models. It weighs less than an ounce.

PHOTO: BOB ABERLE

manufacturers will not actually state "frame rate" controller in their literature in which case, look under the manufacturers' specifications for "switching rate." If you see a figure like 50 Hz (Hertz), it is a frame or low rate type controller. These type controllers are characterized by their small size and light weight. They do provide a proportional speed control through the full range of low to high (full) motor speeds. The only problem is that in the mid-speed ranges, their efficiency drops off markedly. At the top end (full speed) the efficiency is acceptable.

I use a frame rate controller in my electric powered Old Timer model which has won a dozen or more first places for me over the last three years. Flying that type of event involves the application of full power for a specified motor run period, after which the motor is shut down and the model becomes a glider for the remainder of the flight. The reason I use this type controller is because it can handle upwards of 35 to 40 amps, it has a BEC, and it weighs only 1.2 ounces with connectors. For my specific application, this controller is perfect because it was essential to have a minimum weight model.

However, if I had been flying a sport or sport scale model where I would normally be cruising at a mid-range throttle setting, the frame rate controller would have been the wrong choice from an efficiency standpoint. Several very excellent frame rate speed controllers come to mind which are manufactured by Astro Flight Inc., e.g., their Models 215 and 217.

Frame rate controllers in many instances require that a special diode and

Frame rate controllers usually require that you add both a diode and a capacitor across the motor terminals.

PHOTO: BOB ABERLE

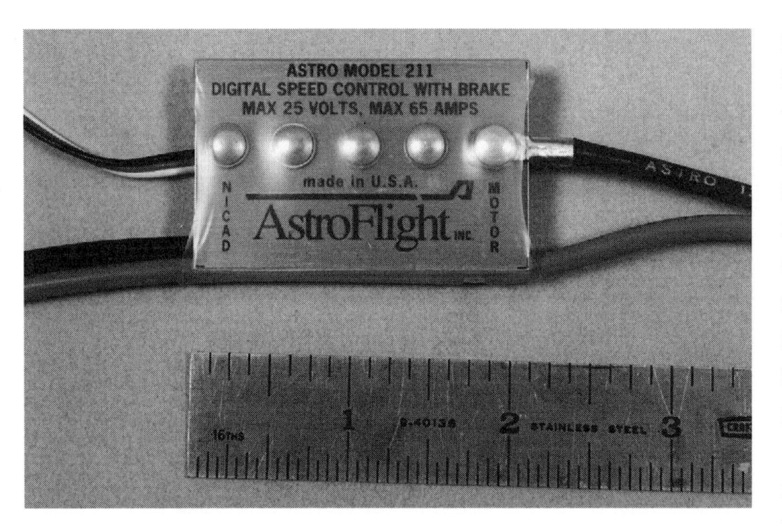

Though small, the Astro Flight #211 high rate controller will take from one to sixteen cells at currents up to 65 amps. It weighs only 1.2 ounces.

PHOTO: BOB ABERLE

electrolytic capacitor be placed across the electric motor terminals. The two Astro Flight controls referenced above require the addition of those two components. They are, in fact, supplied with the controllers along with suitable instructions for their installation.

On some other brands of frame rate controllers, it is possible to have these components installed as an integral part of the unit. I caution you to make sure both of these components are installed, if required, by the manufacturer. Also, be careful, when moving your frame rate controller from one model to another to remember to place those components, if required, across the motor terminal in the new installation, especially if a new motor is being used.

One final point to mention with regard to frame rate or low rate controllers is that they are generally the least expensive of all the types.

High rate controllers are usually advertised as such because the manufacturers want you to know that it is a very efficient speed controller throughout the full speed range. The switching rate of a high rate controller is in the range of 2500 to 4000 Hz. In addition, it is common with high rate controllers to hear a high pitch tone or sound when the motor is turned on, and as the speed is advanced.

High rate controllers are available from many sources and they all work well. They will, in general, cost more than frame rate controllers (the price range can vary from $50 to $200). It is advised to keep in mind your specific speed control application, since that will dictate the cost of the unit you are buying. Two considerations

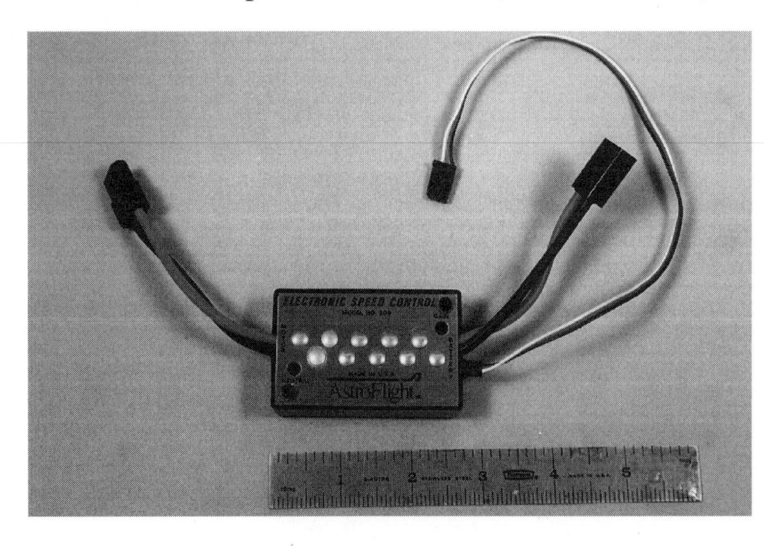

Another Astro Flight high rate controller, the #205, takes six to 40 cells, and currents up to 50 amps.

PHOTO: BOB ABERLE

are the number of cells in your battery pack (to establish the voltage) and the amount of motor current you will be controlling. A small size controller will handle up to 10 cells and approximately 25 amps of motor current. On the other end of the scale are the controllers for the largest of motors which might operate on up to 40 cells at 50 amps current. Be aware that controllers that can handle a lot of voltage and current are also physically large in size.

Many of these high rate controllers offer other features which may enhance your specific application. A battery eliminator circuit (BEC) eliminates the need for a separate radio system battery by allowing you to share power from the motor battery. There are some who do not favor this BEC approach because they fear that the almost fully discharged motor battery will not have enough remaining power to operate the radio for the desired time. My personal experience has indicated the ability to fly for almost an hour after motor battery shut down while flying a competition electric glider and still having the radio system work flawlessly.

One consideration when using BEC type circuits is that you should limit operation to between six and ten battery cells. Below six cells you will not have enough voltage coming out of the regulator to operate the radio and above ten cells you will burn up too much power trying to reduce the voltage to a workable amount for your R/C system.

Another feature to look for in a controller is a built-in electrical brake circuit. If you use a folding prop on your specific model, a brake will guarantee that the motor fully stops so that the prop blades can fold back up against the sides of the fuselage. As mentioned in an earlier chapter, it is not advisable to use a

Jomar's controllers generally use optical coupling to reduce radio interference problems. Both of these controllers are high rate, for a full range of proportional motor speed control.

PHOTO: BOB ABERLE

With a microcomputer as part of its circuitry, the AI/Robotics FX-35D controller performs quite a few tasks. It handles six to twenty cells, and up to 50 amps current. It has a B.E.C., a brake, a soft start, and can even recharge your receiver battery in flight.

PHOTO: BOB ABERLE

motor brake in conjunction with gear or belt drives. Similarly, you might consider a slow start feature that allows the prop blades to first extend before the motor gets up to full speed (I have seen one folding prop almost cut a wing off simply because it began to rotate too fast, too soon).

Included in some controllers and offered as an option on others is optical coupling. Since speed controllers have a direct connection to your R/C receiver, it is possible, under some circumstances, for electrical noise to be introduced into the receiver. To act as a buffer and prevent this type of interference from getting to your receiver, some manufacturers employ an optical coupling device.

About half of my personal controllers employ this feature, yet I haven't experienced any noise interference problems in any of my systems. My personal feeling is that if the optical coupling is offered as an option, you should take it. However, don't allow the lack of that feature to prevent you from buying a particular brand controller.

In the field of high rate electric motor speed controllers some good suppliers are: AI Robotics Inc., Astro Flight Inc., Flightec, Gordon Tarling, Jomar Products, Lofty Pursuits, and Slegers International, who import the line of Sommerauer and Buholzei controllers from Germany. Most of these suppliers offer a full range of controllers that will handle everything from the smallest of models on up to ¼-scale aircraft powered by large single motors and, in some cases, multi-motors. You will have to pick and choose among these suppliers since some offer a BEC while others do not. This also applies to items such as a brake, soft start and optical coupling. The primary criteria should be to buy the controller for the size motor and battery you expect to control. Stated simply, small motors and batteries need only a corresponding small speed controller.

Interest has been increasing for the high rate controller with the added feature of including a micro computer circuit. This new breed of very sophisticated speed controllers can perform every function described under the regular high rate controllers. The addition of the micro computer permits the inclusion of some very worthwhile features. Probably the most important added feature involves safety.

As pointed out earlier under the disadvantages of electric power is the fact that accidentally turning a motor on can be dangerous (any time power is applied to an electric motor it runs!). Unlike an internal combustion engine, the electric motor doesn't need a prime, a lighted glow plug, or the prop flipped to get it started. Therefore, accidental electric motor start-ups must be watched carefully.

A safety tip: take the prop off the motor shaft when installing a new motor system in your model, or even when doing any work on your electric powered model. An electric motor, without a prop attached, is safe.

Getting back to our micro computer speed controller: what several manufacturers have come up with is a smart circuit that only allows the motor to run if certain parameters have been met.

For example, if you turn on your R/C system and motor arming switch (if you have one), but the R/C transmitter is still turned off, the motor will not run at all. That leads to the classic problem: you turn both the plane system and the transmitter on, but fail to realize that the throttle control stick was in the up or full power position. On most controllers that situation would cause the motor to spring to life. If you weren't prepared for that situation your model could take off through the pit area and possibly hit and/or injure someone.

With the micro computer circuit in the controller, the motor will not turn on. In fact, to make the motor run, you first have to move the transmitter throttle stick down to the off position and wait a few seconds. Then, and *only* then when you advance the control stick, will the motor begin to run. I personally won't be happy until all my controllers have this feature.

Some micro computer controllers offer even more features. One in particular, the FX-35D offered by AI Robotics Inc., includes such additional items as active overload protection, thermal protection, power-on safety delay and a choice of soft start rates. The FX-35D also has a patented feature called a Sequential Arming Switch (or SAS). This is a master three position switch which you mount on the side of your fuselage. The first position is all-off. The middle position activates only your R/C system enabling you to check out your control surface movement (the motor system is still disarmed). Only when you advance the SAS to the last position is both the radio on and the motor circuit armed.

The final item on the FX-35D controller is an in-flight battery charging capability. To further explain: let's say you have a large model operating on 18 cells which means you can't use a BEC, but you still want to keep your weight to the minimum. All you need to power your R/C system is a very small four-cell Ni-Cd battery pack, such as 150 or 225 mAh capacity. You can use a small, lightweight pack because if, for any reason, the voltage of that pack drops to 4.6 volts during a flight, the battery will be automatically recharged by the motor battery. This is a clever application which provides an in-flight recharging capability of the R/C system battery, if needed.

In addition to the micro computer controller offered by AI Robotics, there are several other sources such as ACE R/C Inc., Astro Flight Inc., Flightec, Gordon Tarling, Jomar Products, and Lofty Pursuits who each offer a variety of special features.

The last controller to mention is the special or dedicated type necessary to operate the new breed of brushless DC motors. Because these controllers are so specialized you must buy the companion controller for each motor (you can't mix or match between vendors). These controllers, in general, are the largest and heaviest of all the controllers, some weighing as much as 3.5 ounces, but they do offer many of the optional features already discussed in this chapter.

At the present time, while these brushless DC motor systems are heavier and more expensive than other systems, it is my feeling that in time this new technology will improve. I see the motors and their companion controllers getting lighter in weight, smaller in size and their prices coming down. The key advantages of a brushless DC motor system are improved reliability, long service life and the claim of higher efficiency. At the present time there are three sources for brushless DC motors and controllers: Aveox Inc., Flightec, and MaxCim.

One last thought on controllers in general: similar to the motors and the batteries, speed controllers also generate heat. The first generation of controllers years ago required add-on heat sink plates that looked much like a radiator to help dissipate the heat generated. Modern semi-conductor technology has made all of these devices "thermally" more friendly for our application. However, it is still a good idea to get some air circulating around your speed controller wherever it is mounted in your fuselage, and try to keep it a distance away from the motor and battery, but this is not a major concern today.

connectors & wires

Connectors and wire are the two items which allow you to "hook up" or connect the entire electric power system. In the simplest form, and in very small size electric powered models, you could get away without using any connectors (e. g., the motor, speed control or some form of switch, and the battery could be permanently wired together). This was the way the basic system was portrayed earlier in **Chapter 3.** Charging the battery could be accomplished by attaching clips leads to the battery terminals. The resulting system would benefit because there would be no electrical losses (due to the use of connectors) in the circuit.

On the other hand, you wouldn't be able to substitute battery packs or remove the battery from the model for charging purposes. Most modelers will also want to consider the use of connectors for the flexibility of substituting components and for maintenance purposes (e. g., removing a motor to replace the brushes, shaft and/or armature, etc.). **Figure 8-1** provides an expanded system block diagram showing the use of connectors.

If for no other reason it still makes sense to have at least one set of connectors at your battery pack so that the power source may be disconnected. This is important when you consider transporting or storing your model aircraft. Even when using such things as fuses and master arming switches, you will never have any safety problem associated with electric power, if your battery is physically disconnected from the circuit.

Figure 8-1 Block diagram of an electric powered flight system showing use of connectors.

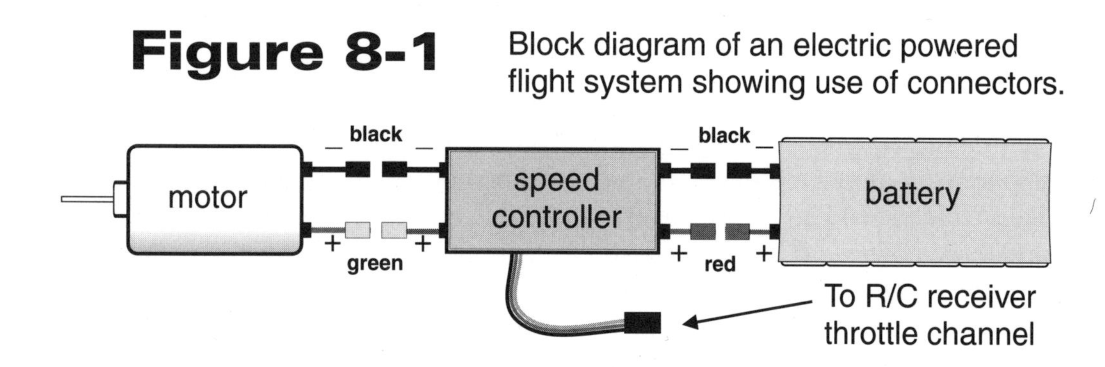

The use of any electrical connectors

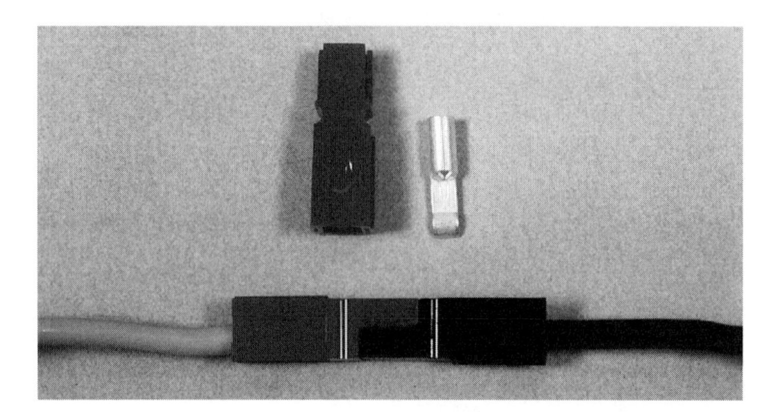

Probably the most popular connector used in electric flight is the Sermos R/C Snap Connectors. They come in red and black plus other colors. The wire is soldered into the silver plated pin, and the tongue/wire inserted into the Lexan housing. This snaps into another Sermos Connector.

PHOTO: BOB ABERLE

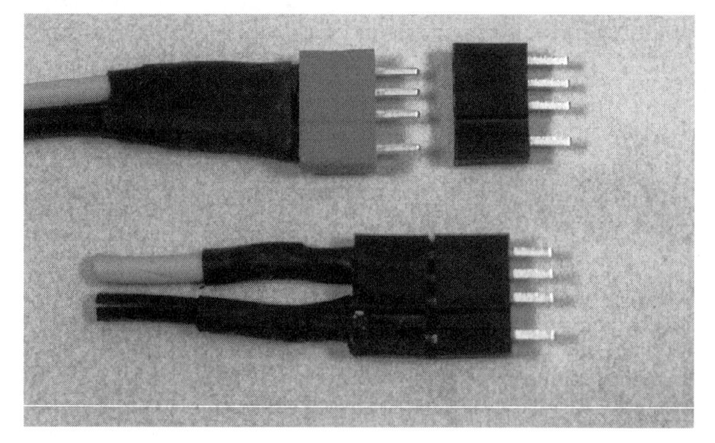

The Deans four-pin connector can be used for smaller models, with lower current demands. In this case each wire is soldered to two of the pins to reduce current loss. Note the slight gap between the one pin and group of three above it. This helps insure correct polarity.

PHOTO: BOB ABERLE

will cause a certain loss of power in your system, but depending on your choice you can minimize those losses. Over the years one very popular connector has emerged which seems to be the favorite of the electric powered flight enthusiasts. These connectors are made by the Anderson Power Pole company and were introduced into our hobby by a Connecticut modeler, John Sermos. John has marketed these wonderful connectors under the name Sermos R/C Snap Connectors and thus the name "Sermos Connectors" has caught on throughout the hobby.

The Sermos connector consists of a molded plastic case with an internal silver plated contact pin. The case molding has been cleverly designed with grooves and strips to allow these connectors to be joined into pairs. The joining can also be done in a way that allows you to establish polarization. In other words the connector pairs will plug in one way but not another.

Additional polarization assistance is offered in the choice of several connector colors. You can obtain the usual red and black colors to denote the positive and negative battery terminals. In more recent times, modelers have been using green Sermos connectors on the motor positive terminal to distinguish it from the battery connection. I have found this helpful since, on several occasions, I have inadvertently attempted to charge my speed control rather than the battery. That situation could have easily blown the speed control had I not discovered my problem immediately. The use of different colored Sermos connectors is a good idea.

Basic assembly of a Sermos connector involves first attaching the wire to the silver plated pin, after which the pin/wire assembly is inserted into the plastic casing until a "snap" type sound is heard. At that point the pin is locked in place. There is a tool that Sermos sells that allows you to remove these pins should the need ever arise.

Attaching the wire to the Sermos pin can be done in two ways. Equipment manufacturers can buy a special crimping tool from the Anderson company at a cost of approximately $130. With this tool, a wire can be mechanically attached to a pin and a perfect connection is quickly made. Since we all can't pay $130 for such a tool, the average modeler must resort to soldering the wire to the pin.

To do this job correctly, you first need the correct soldering iron. I use the Ungar pencil type iron which consists of an insulated handle, heating element, and special tip. Both the wattage of the heat element and the choice of tip is important to this particular application so let me specify them exactly. The insulated handle is an Ungar No. 7760, the heating element is an Ungar No. 1237S (33 watt thread-on type) and the tip is an Ungar PL-113 (¼ inch thread, with a ⅛ inch chisel tip).

soldering Sermos connectors

1

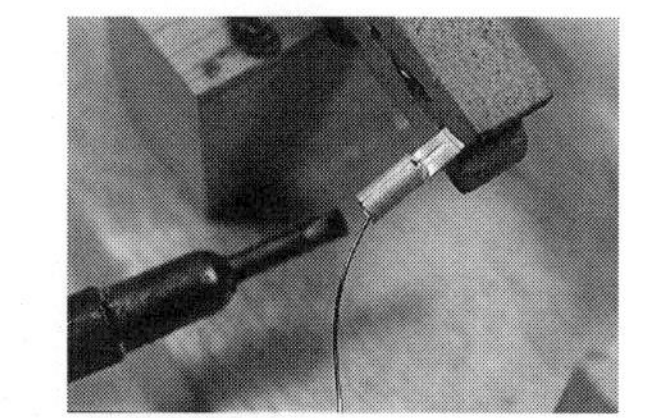

Hold the silver plated pin with a clothes pin, insert solder, apply heat, and melt solder in the pin pot.

2

Keep the heat applied and the solder molten while you push the wire into the pot. Allow it to cool.

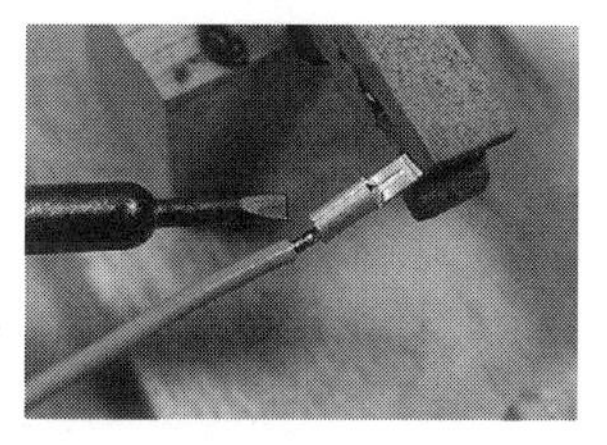

3

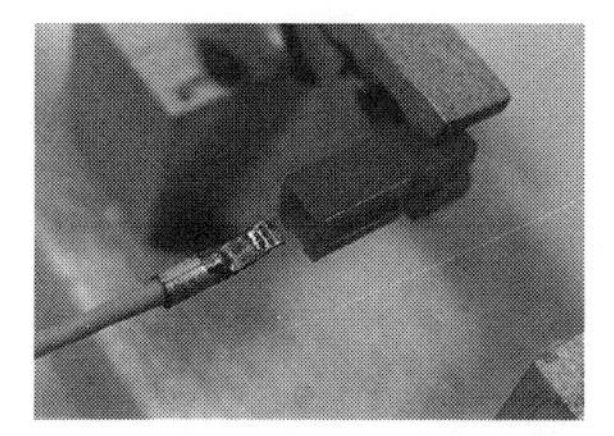

Select a plastic housing that matches the wire insulation color and push the tongue into the back of the housing until you hear a click.

Many local electrical supply stores carry this line of soldering equipment so you might try them as a source of supply. For the solder itself I always use the Radio Shack No. 64-005 which is a 60/40 composition with an .032 inch diameter.

For the actual assembly I made up a small holding fixture consisting of a wood clothes pin and a small block of wood. I grip the front end of the Sermos pin with the clothespin, leaving the rear end with the solder cup open. About ¼ inch of the insulation is stripped off the end of the wire exposing the copper strands. Next I hold the soldering iron under the Sermos pin and start the heating process. After a few seconds I introduce the solder up into the "cup" portion of the pin.

Another type of connector has recently appeared, the Astro Flight "Zero Loss" type with gold plated pins. These are easy to assemble.

PHOTO: BOB ABERLE

Keep feeding in the solder and allow it to melt until the cup is filled. Hold the soldering iron tip in place, then take the wire and insert the bare stranded copper end up into the Sermos pin as far as it will go. At that point, hold the wire in place and remove the soldering iron, allowing the solder to cool. Don't rush this step since it might take about 10 seconds for the solder to cool sufficiently. Clean off any excess solder flux with the help of acetone and an old tooth brush.

A caution: do not use any extra rosin type solder flux in this process. The only flux necessary is that which is built directly into the solder. If you use additional rosin flux, there is a good chance that the solder will "wick" back up the copper wire and inside the insulated covering or jacket. This wicking won't be obvious initially; however, later on you will discover a very brittle point on your wire just beyond the end of the connector casing. This can become a fatigue point where the wire might easily break off at some later time.

After the wire is installed in the pin the only thing remaining is to insert the pin into the housing. Sermos supplies an excellent set of instructions in this regard. If you insert the pin correctly you will hear a "snap" sound when the pin is finally in place. If you don't hear that snap you probably did something wrong. After completing a Sermos connector assembly try mating it with another assembled Sermos connector. When you do this you should again hear a "snap" sound. If you don't at this point, you made a mistake somewhere and it would be best to backtrack to investigate.

If Sermos connectors don't snap, you will certainly have a problem. I've had motors fail to start simply because I improperly installed a Sermos pin. You may have trouble the first few times, but I can assure you that you will learn how to install Sermos connectors quickly and reliably.

One final point to note when using Sermos connectors: they have two different types of plastic housing material. The standard connector is made of a Lexan material, which is not impervious to any hydrocarbons. If you were to spray a contact cleaner or an accelerator chemical for CyA glues near these connector cases, they would literally crumble into a powder in a short time. If you consider this a problem it is advised that you spend a little more and get the new Sermos Super Connector, because nothing will attack this new material.

To distinguish the standard from the new material Sermos is using different colors and has varied the type of finish. The new "Super" type will be a tangerine color rather than red and the finish will be a flat or matt type as opposed to the standard material, which has a shiny finish. Be sure to stipulate the type you want when placing your order.

There is another excellent connector for our electric power use that has come on the market more recently. It is the Zero-Loss type manufactured by Astro Flight Inc. This makes Bob Boucher of Astro Flight a total supplier in the sense that he manufactures motors, speed controls, and chargers, and now he has his own connectors as well.

I won't dwell on the assembly of the Astro Flight Zero Loss connectors because the procedure is obvious. The installation instructions supplied are excellent in every regard. Again the wire end is stripped, the gold plated pin (in this

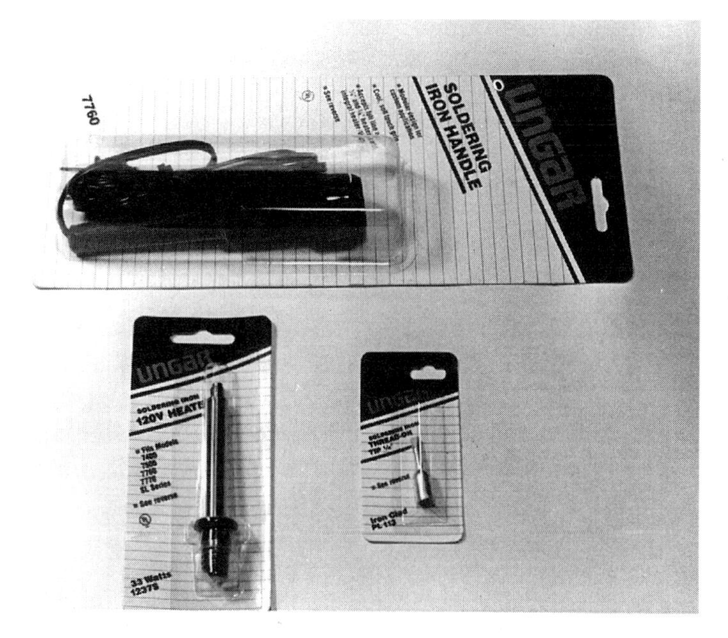

This soldering equipment is perfectly suited for all electric flight needs. It consists of a soldering handle that accepts a 33-watt element, and a small, screw-on spade tip. The heat range of the iron shouldn't exceed 40 watts.

PHOTO: BOB ABERLE

case) is held in a fixture while the hollowed out inside is filled with solder. With the solder still molten, the wire is inserted all the way into the cup. Let the connection cool and then simply pull the pin inside the black plastic housing. The Zero Loss connector housings are all black in color. Polarity is established by mechanical means (i.e., the housing is "keyed" so that the male and female connetors can only mate one way).

One other type of connector, manufactured by the Deans Company, might see some service in the smaller size models where the current requirements wouldn't be much in excess of 10 amps. A popular way of using the Deans connector sets is to use the four-pin variety and connect two pins together for each wire (two pins for the positive wire and the other two for the negative). The Deans pins can be soldered very easily. It is a good idea to cover the exposed connection end with heat shrink tubing to prevent any possible short circuits. Deans connectors are both small and light.

The wire selected for an electric motor system is not that obviously important, but it is. The primary concern should be the current carrying capability. Also important is the wire flexibility and the number of strands that make up the wire. The more strands, the better. Don't attempt to use something like household "zip cord."

Wire size in the U.S. is stated as a gauge or number. The lightest gauge wire we would use is #16, which would be for the smaller motor systems drawing 10 to 15 amps, like the HiLine Imp-30, Elf 50, Peck Polymer Silver Streak, Graupner Speed 400, Kyosho AP-29, etc.

The most popular size wire for general use is the #14 gauge. This can handle the more usual 20-25 amps current drain and up to 10 to 14 battery cells. Several suppliers offer #13 gauge which is just a little heavier.

The heaviest gauge for our electric powered use is the #12. This would be employed for the "big stuff" like 40 and 50 amps for certain sprint type competition flying as well as the large motor systems employing battery packs with 20 to 36 cells. As a point of information, #16 gauge would be used with the Deans connectors. The Sermos pin can accept #12 gauge wire, but it is a tight fit. Astro Flight's Zero Loss connectors are claimed to be able to use #13 and #14 gauge. I'm not sure if they will accept #12.

SR Batteries Inc. offers #16, #14 and #12 gauge with yellow and black colors. The yellow color is intended for positive polarity. SR did this right from the start of their business as a point of marketing identification. Anytime you see yellow and black wiring you know SR Batteries is involved. Their wire is good, and most important, the plastic sleeving does not melt easily at normal solder temperatures.

Jomar Products offers their Super-Flex wire in both #15 and #12 gauge. Other sources of wire include manufacturers like ACE R/C Inc. and Astro Flight Inc.

fuses, switches & charging

The addition of a fuse in your electric power flight system can help prevent damage to the aircraft wiring and its components. It might even prevent excessive heat build-up from turning into a fire. A stalled electric motor can cause this kind of heat build-up.

For example, you roll out after a landing and the plane noses down. You fail to realize that the transmitter throttle stick is still in the on position (the motor wants to run, but the prop is touching the ground and can't turn). The result is a stalled motor that begins to overheat. The same problem could occur if a component failed inside your speed controller.

In either situation, the use of a fuse offers positive protection for the entire electrical system, because when the current rating of the fuse is exceeded it melts and opens. This, in turn, opens the electrical circuit which effectively disarms the entire system. The fuse in this case provides the same kind of protection that a fuse or circuit breaker provides for your household electrical wiring.

Most model flying clubs stipulate the use of protective fuses in electric powered aircraft as a general safety requirement. More and more contest directors are also making the use of fuses mandatory for competition flying. It is true that certain losses do result from the use of a fuse, but if everyone used them, everything would be equal.

The spade fuse (bottom) is preferred in electric applications. Used with a special set of Sermos Connectors, it can function not only as a removable fuse, but also as the arming switch.

PHOTO: BOB ABERLE

Common sense might tell you that a fuse can be placed almost anywhere in an electric circuit. When the fuse is overloaded and melts, the circuit opens and the various components are protected. However, it isn't quite that simple for our application of electric powered flight.

The problem is the frequent use of the BEC (battery eliminator circuit) in electric power systems. As discussed in **Chapter 7**, some speed controllers have a BEC circuit that allows the radio system to share the same motor battery. Depending on where you place your fuse in the system, it would be possible for all power to be cut off, which would render

Figure 9-1 Locating a fuse in an electric powered flight system when using a B.E.C.

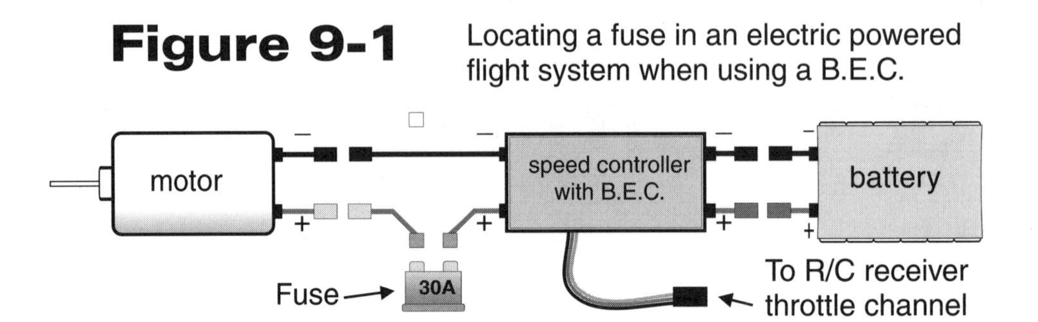

the BEC useless and the radio control system as dead as the motor system.

Figure 9-1 shows the recommended location for a fuse when a BEC circuit is being used. You will notice, in this situation, that the fuse is placed close to the motor, and away from the battery. If the motor overloads and the fuse blows, the battery will continue to power the speed controller whose BEC will keep the radio system operating, and let you safely land your aircraft. The only problem with this type circuit is that, if there is a failure in your speed control, the fuse may not help. If that is the case, and the speed controller fails, so will the BEC circuit. In that event, the radio will go dead and the aircraft cannot be controlled.

On the other hand, if your radio system is being powered by a separate battery pack, you will want your fuse protection close to the main battery. **Figure 9-2** provides a block diagram of such a circuit. In this instance both the motor and speed controller are protected. If the fuse blows, the entire system is shut down and protected, yet the radio will continue to operate on its own separate battery pack.

Trusting that you have been convinced to use a fuse, what type should you select? My choice is the flat or spade type automotive fuse that is available in currents up to at least 30 amps. For general flying, 20- or 25-amp rated fuses are adequate. The 30-amp fuse can actually take

Figure 9-2 Location of the fuse in a power system when ***not*** using a B.E.C. (receiver uses a separate battery)

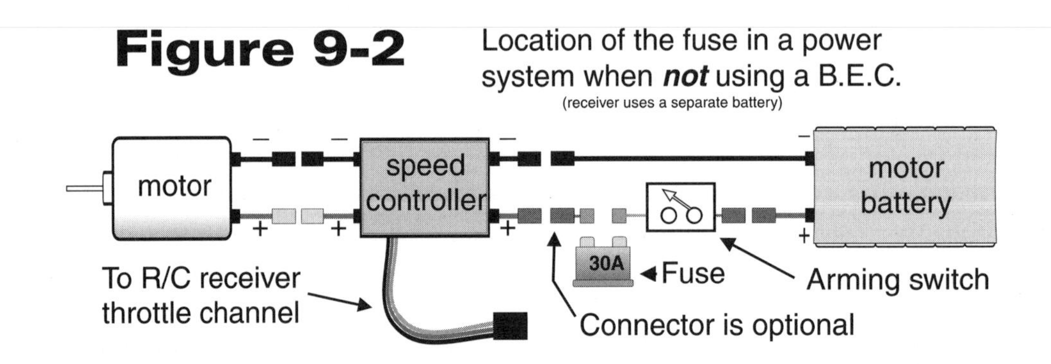

a load of 40 amps for a period of 30 minutes before blowing. That should give you a rough guide when it comes to selecting the current rating of a fuse for your particular application.

Until recently, these fuses were generally soldered directly into the circuit because there were no commercially available sockets for them. Sermos has now come out with a modification to their basic snap connector, which accepts a spade fuse and can also be mounted to a fuselage side, with the fuse projecting to the outside.

Another benefit to using an externally mounted fuse is that the same fuse can also be used as an arming switch. Even though many of the new speed controllers on the market have the built-in safety feature that prevents accidental motor start ups, a separate arming switch is still a good idea, even if redundant. By simply pulling the fuse out of its Sermos holder, your electrical system is totally disabled. This one fuse therefore, prevents both circuit overloads and accidental or inadvertent motor start-ups.

If you still would like to have a separate arming switch and fuse (see **Figure 9-2**), the best choice would be a toggle or rocker type switch. Generally, this switch would only be turned on or off when the motor system was off and not drawing any current. In this case a switch with a current rating of 5 to 10 amps is sufficient for most applications, even though the motor itself might draw 30 amps when running. To add extra current capacity to the switch, it is a good idea to buy the DPDT variety and then connect across both sets of terminals to increase the effective current (see **Figure 9-3**).

Figure 9-3
Toggle, or rocker type arming switch.

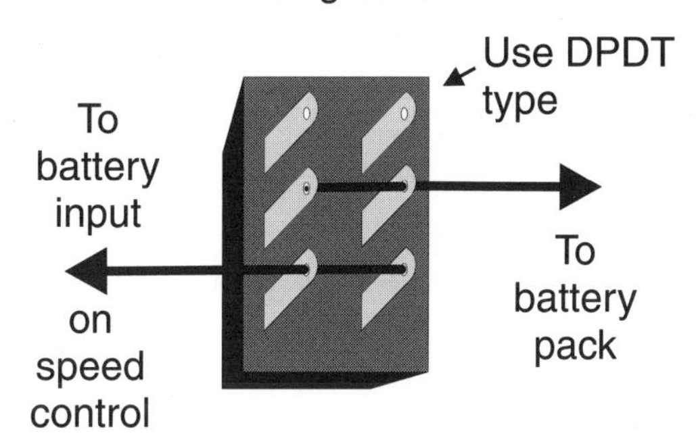

Bridge terminals to enhance current carrying capability

Another item you might want to consider in your electric power flight system is a battery charging jack. Many electric flyers prefer to unplug and remove their battery pack from the aircraft for charging purposes. It gives the battery a chance to cool down which is important. You can also substitute a freshly charged pack, while the depleted one is placed back on the charger.

Some battery installations make it necessary to remove a hatch cover or even the entire wing to gain access. In some circumstances, it may prove easier to charge the battery while it remains inside the aircraft. If that is the case, you might consider adding a pair of charging jacks which easily allow the battery to be

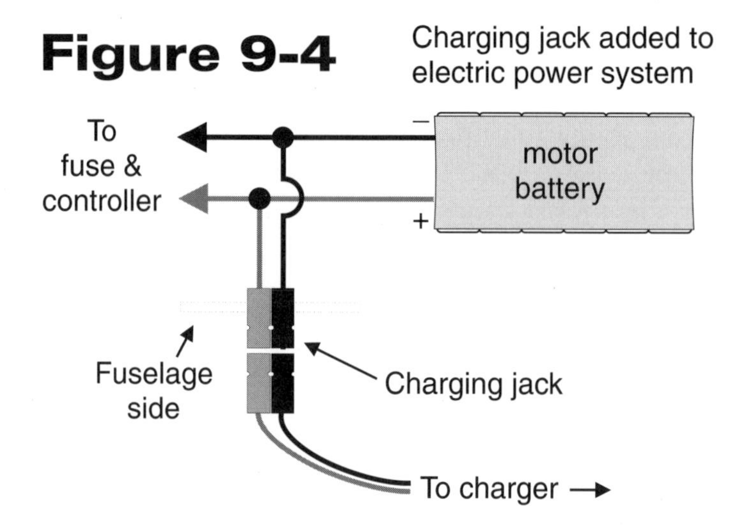

charged without dismantling the model.

Again, a modified Sermos connector set can handle the job. It will look very similar to the Sermos fuseholder, except that it accepts a mating half of a regular Sermos connector. The Sermos charging jack assembly can also be mounted on a fuselage side where it is easily accessed by the charger cable. The key point to remember in using a charging jack is that it must be attached as close to the battery as possible (see **Figure 9-4**).

radio systems

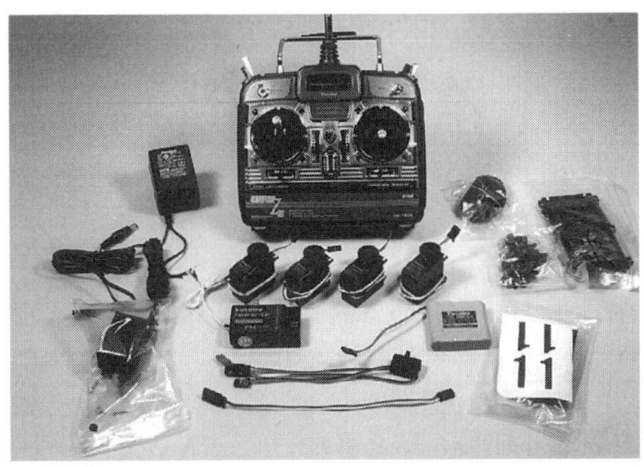

Though a good inexpensive radio, the Futaba 6-channel Skysport 6VA represents a common problem for electric models. For smaller models, you will have to substitute a lighter receiver and servos.

PHOTO: BOB ABERLE

Explaining the operation of a radio control system would be very lenghty, and in all honesty the instruction manuals provided with these systems do an excellent job of explaining both installation and operation. An excellent reference on R/C systems is George Steiner's book, *The A to Z Radio Control Electronics Journal*. What I want to do in this chapter is highlight some of the items that will have the most value to those of you pursuing electric powered flight.

The most important thing to consider when planning an electric powered model aircraft is to keep the weight to a minimum. When you take into account the weight of the motor and battery pack, you already have a considerable mass to contend with. That leaves a modest weight allowance for the model structure and radio control system. You have to build the model light, yet strong enough to support the gross weight of the aircraft. The airborne R/C system components must also be as light as possible and still be able to reliably and safely control the model in flight.

The problem is that most R/C systems are usually supplied with average size components (receiver, servos, battery pack, and switch harness). What you really want is a system which has small size and thus, lighter components. In many cases, these smaller components must be purchased separately, at additional cost.

In an earlier chapter, mention was made of the Futaba Model 4NBF Conquest four-channel R/C system. There is also a new six-channel version of this same system that contains some added control features and is known as their Model 6VA SkySport. Both of these are wonderful R/C systems, but they are supplied with the standard size Futaba R127DF receiver which weighs 1.5 ounces. My preference for electric flight would be to use the miniature Futaba R-148DF receiver which weighs just 1.0 ounce.

All of these receivers are dual conversion but the bottom two, a Futaba R-148 (L) and an Airtronics Micro which both weigh one ounce, would be better suited for electrics.

PHOTO: BOB ABERLE

The usual servos supplied with Futaba radio systems are their S-148 at 1.5 ounces each. My favorite, for electric flight, is their tiny, yet powerful, Futaba S-133 servo that weighs only 0.6 ounce each. Even the normal 500 mAh airborne battery packs can be replaced by their 250 mAh packs at half the weight.

One of the few instances where you can buy a radio system that does come with a smaller receiver and servos is the Futaba Model 7UGFS, which is known as their Super Seven Sailplane radio. This, however, is not a basic radio, so will cost a little more. I not only like this radio because of the small size of the airborne components, but because the microcomputer transmitter can operate up to four different model aircraft. When you want to control your second aircraft, all you have to do is purchase more airborne components, not a full system.

The other nice feature of this sailplane radio is that it offers special control features such as, flaperons, spoilerons, and crow. All of these extra features come in very handy when trying to fly an electric powered sailplane. I might add that this same sailplane radio still has all the control features necessary to fly all of the other types of fixed wing aircraft as well.

As you continue in electric powered flight, you might want to consider electric powered helicopters. Should you attempt helicopters, please be advised that a specialized radio control system will be required which is dedicated to this type of flying.

Relatively inexpensive radio systems are now on the market with micro computer control. Both the JR XF-622 and the new Hitec RCD Flash provide two-

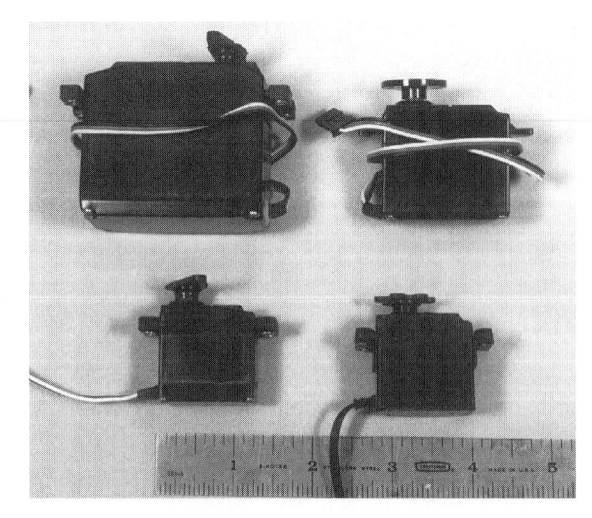

Since there can be two, and probably more, servos in an R/C electric model, good servo selection can reduce weight. This range of servos shows what's commonly available, with a small servo like the 0.6-ounce Futaba S-133 at the lower right more preferred.

PHOTO: BOB ABERLE

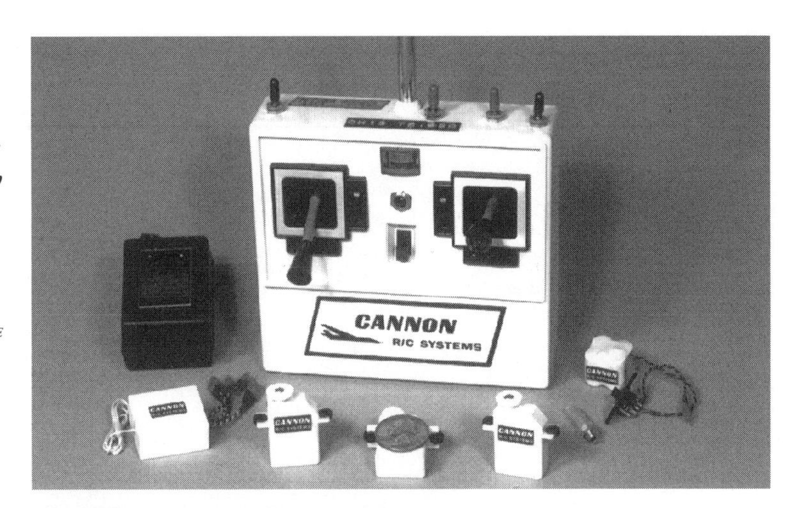

For really small electrics and indoor R/C, the new Cannon Ultra-Micro radio system is their smallest yet. The complete 3-channel airborne system with receiver, three servos, switch harness, and 4-cell 50mAh battery is just 2.0 ounces.

PHOTO: BOB ABERLE

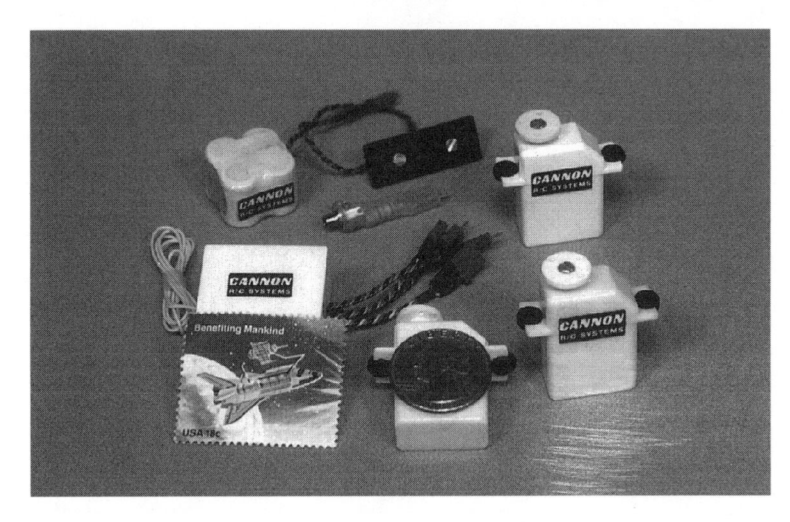

A quarter graphically shows the size of a servo in the Cannon Ultra-Micro system, while a postage stamp does the same for the receiver. The receiver weighs 0.41 ounce, and each servo 0.30 ounce.

PHOTO: BOB ABERLE

plane memory circuits, which means that one transmitter can control two separate aircraft, provided of course, that you purchase the same compatible airborne components. Both of these manufacturers also have complete lines of light weight airborne components (receivers and servos) for you to choose from.

One company in particular over the years has specialized in what must be considered the smallest of R/C systems. That company, Cannon R/C Systems Inc., has recently completed development and is now shipping what they call their, "Ultra Micro" R/C system. The tiny servos for this system weigh approximately ¼ ounce. A complete two channel airborne system, including a receiver, two servos, switch harness, and a four-cell 50 mAh battery pack weigh 1.6 ounces. A radio system such as this should be considered for the smallest of electric powered models and especially for indoor R/C flying.

Any of the following manufacturers can provide radio control systems to operate electric powered model aircraft:

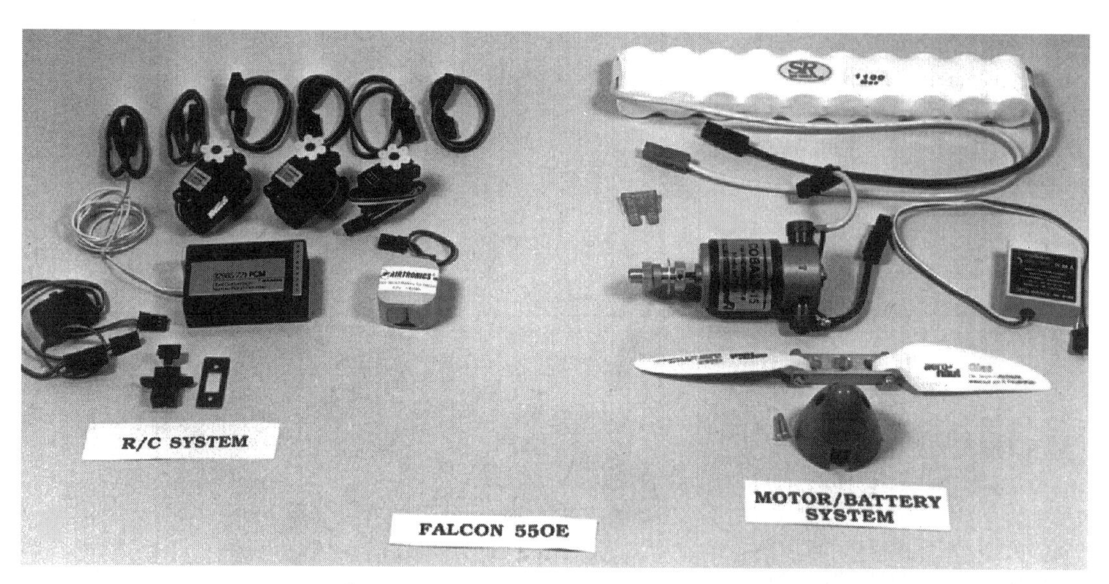

All this equipment is on board the author's Airtronics Falcon 550E electric sailplane and represents what goes into an electric-powered model. Total weight of airframe, radio system, and electric power system comes to 50 ounces.

PHOTO: BOB ABERLE

ACE R/C Inc., Airtronics Inc., Cannon ACE R/C Systems Inc., D.A.D. Inc., Futaba Corp. of America, Hitec RCD Inc., JR Remote Control distributed by Horizon Hobby Distri- butors, and Polks Modelcraft Hobbies Inc. Addresses will be found in the appendix so that you can write for brochures, catalogs and price listings.

Radio control systems operate on discrete channels or frequencies assigned by the Federal Communications Commission (FCC). The modeling community and the Academy of Model Aeronautics (AMA), in cooperation with the FCC, has come up with a system of frequency or channel identification which allows for non-interference operation with other modelers at any local flying field. The identification system involves primarily the use of channel number flags or placards that are attached to the transmitter antennas. When you arrive at your flying field to test and/or fly your model the first time make sure you find out about the R/C frequency control system, as well as the safety rules, for your particular flying site. This will be discussed again later in **Chapter 14** on Flying Electrics.

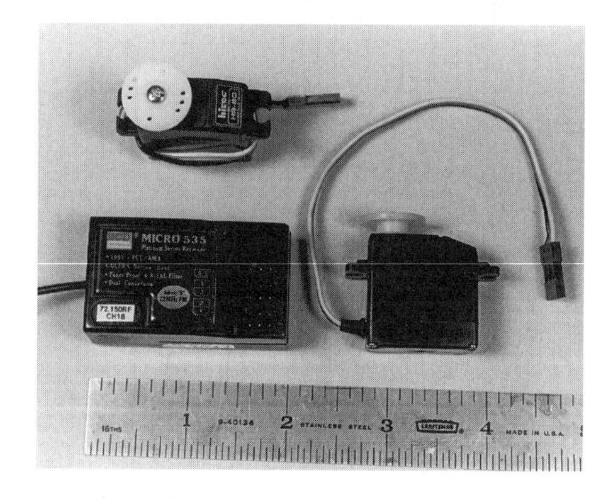

This combination of Hitec RCD radio gear would be an excellent selection for electrics. It shows the Micro 535 receiver that weighs 0.9 ounce, and two HS-80 micro servos that each weigh 0.6 ounce.

PHOTO: BOB ABERLE

selecting powertrains

The Airtronics Falcon 550E represents one choice of powertrain options available to electric powered flight. A direct drive Astro Cobalt 15, with ten 1100 mAh cells gives a burst of energy for a 30-second motor turning the prop at 14,000 rpm. That draws 42 amps current.

PHOTO: BOB ABERLE

After having become familiar with all the basics of electric powered flight and the various components, the final step is one of application. How do you select the specific motor, type of drive (direct, gear, or belt), propeller, and battery for your particular model? The wrong choice, or even a poor choice, of electrical components has disappointed many an aspiring electric flyer.

The first step in the selection process is for you to decide on the particular "mission" for your electric powered model. By that I mean what do you intend for your model? Will it be an endurance model, like a powered sailplane, or a scale model of some full size aircraft? Will it possibly be a speed model, such as a pylon racer, or a basic fun-fly or sport model that, as the name implies, simply provides fun?

Maybe it will be a seaplane or a model with floats that can take off and land on water. You might even want to consider special situations such as multi-motors (twins and the like) or electric ducted fan power for simulated jet aircraft operation.

Once you have made the initial choice of the mission you will undoubtedly have a model in mind. The next step in the se-

Modelair-Tech's Beechcraft D-18 shows another way to develop an electric powertrain. It uses a single electric motor inside the fuselage that turns two belt drives that go out to each nacelle and prop. Flight performance is excellent.

PHOTO: BOB ABERLE

lection process is the form of motor drive to use. In other words, will you require direct motor drive, a gear reduction drive, or a belt reduction drive. Direct drive (propeller attached directly to the motor shaft so that it turns at the same speed as the motor) is suited generally for fast moving models where speed is important, but duration is not.

Pylon racers and high performance competition motor sailplanes fit this category. Many electric competition models use direct drive. To a degree, some of the very small schoolyard or even indoor models operate well on direct drive.

Gear reduction drives allow the propeller to turn at slower speeds than the motor shaft. The idea is to turn larger diameter props at slower speeds (rpm), requiring less current drain from the battery. This adds up to more thrust and longer motor run duration. Gear drives can be made quite compact in size. They also cause a change in the direction of rotation of the prop. That means the electric motor must have its polarity reversed and that the motor itself must be re-timed for more efficient operation. Gear drives

Direct drive motors are a necessity for models like electric pylon racers, where speed is important and duration less so. The models which Ron Farkas (L), Bob Aberle (C), and Tom Hunt (R) hold are all Speed 600 pylon racers.

PHOTO: BOB ABERLE

Table 11-1

Typical Choices Electric Powered Flight Aircraft/Motor Type/Batteries

Model Type	Wing Area (sq. in.)	Gross Weight (oz.)	Wing Loading (oz./sq.ft.)	Motor	Gear Box	Prop Dia × Pitch	Battery Cells / mAh
Very Small Sport	170	12–14	10–12	HiLine IMP-30	(Direct)	5 × 3	3 / 700
Small Sport	210	18–24	12–17	Kyosho AP-29	(Direct)	6 × 4	5 / 600
Small Scale	300	32	15	Astro Geared 035	2.38/1	10 × 6 (Folder)	5 / 1250
Small Pylon Racer	180	16	13	Speed 400	(Direct)	6 × 3	7 / 500 to 600
Old Timer	630	35	8	Astro FAI Geared 05	2.38/1	12 × 10	7 / 800
Electric Sailplane Class A	650	45	10	Trinity Sapphire Ferrite 17T 05	3.5/1	13 × 8	7 / 1000
Primary Trainer	400	50	18	Astro Cobalt Geared 05	2.38/1	12 × 8 (Folder)	7 / 1100 to 1700
Electric Sailplane Class B	650	47	11	Astro FAI Cobalt 15	(Direct)	9 × 5 (Folder)	10 / 1100
Pattern (Light)	640	84	19	Astro FAI Cobalt 25 Geared	1.63/1	11 × 10	16 / 1500
Pattern (Heavy)	630	112	25	Speed 700 Ferrite	2.5/1	12 × 8	16 / 1500
Large Sport	1400	220	23	Astro Cobalt 40 Geared	1.63/1	14 × 8½ (Folder)	21 / 1700
Large Scale	1000	160	23	Astro Cobalt 60 Geared	1.63/1	16 × 8	24 / 1700

For the competition tasks required of electric powered Old Time models, gear drives provide the best benefit. This is the author with his electric powered Schmaedig Stick.

PHOTO: BOB ABERLE

are excellent choices for slower flying scale models, sport models, trainers, and powered sailplanes. They are also well suited for the heavier models, since they can be set up to develop more thrust than direct drive.

Belt reduction drives produce about the same desired results as gear drives. One of the advantages over gear drives is the fact that the motor need not be reversed and therefore, no re-timing is necessary. If you have a sealed ferrite motor, where adjustment (re-timing) is not possible, the belt drive is the better choice. Belt drives are generally larger in size than gear drives, but can usually handle motors of larger size and power ratings.

Belt drives are excellent for the larger scale models, since they are easily capable of swinging large diameter props. That fact also makes them well suited for slow flying models, sport models, and large size trainers.

The final choice in the selection process is the way the plane will be built. By that I mean building from a kit that was expressly designed for electric power, or building from scratch (using plans that were purchased, or your own design), or converting a plane originally intended for "wet" powered to electric (be it in kit or prefabricated form).

The first choice is obviously the easiest because the designer/manufacturer has already taken the trouble to select the correct motor, type of drive, propeller, and the battery cell count/capacity. The number of such kits available at this writing is somewhat limited, but the number is growing rapidly, especially so now that several excellent gear and belt drive units, along with some large, yet inexpensive motors, are being marketed.

But the current limited number of kits specifically designed for electric power leads to the number two choice, which is to build from scratch (buy your own materials and cut out your own parts) using either purchased plans or by designing the model yourself. In either case the approach is to first pick a motor and battery size for your project. I'll give you some suggestions on this in a moment. After choosing a motor/battery, compute the wing area of your proposed model. From there you can either proceed to design your own model or you may go to one of the many plan services and purchase a set of full size plans that comes close to your needs.

The final option open to an aspiring electric powered flyer is to convert an existing "wet" powered model. This could be a model that was already built and flown by gas power. Or it could be a kit or almost ready-to-fly (ARF) type model that was originally intended for "wet" power. In either event, you would first compute the wing area and then make a good estimate on what the final weight might be.

Obviously certain assumptions must be made during this planning process. Once you have done this you could consult the chart provided in this chapter (**Table 11-1** "Typical Choices Electric Powered Flight Aircraft/Motor Types/Batteries"). This data will certainly get you into the right ball park.

The next step beyond using the chart just referenced is use a computer to determine what you will need. Several enterprising electric power enthusiasts have recently developed computer programs for use on home PCs that will allow you to more precisely select and size your electric flight systems for most model aircraft applications.

The first, and the one I am personally familiar with, is the Electro Flight Design Program developed by Bob and John Kress of Kress Jets Inc. This program is intended for IBM compatible computers (386 or higher) using Microsoft Windows. It consists of four sections as follows: 1) calculation of model drag and required horsepower; 2) electric motor performance; 3) optimum propeller selection; and 4) fixed propeller selection.

In the first section of this program the user inputs such items as model weight, wing area, span, and air density for the altitude and temperature conditions being studied. Under motor performance, you can either enter parameters of your own or you can choose data for 15 different popular motors which are supplied with the program. The remaining two sections take into account the variables associated with both direct drive and gear/belt drives, along with the actual recommended sizing of the prop.

Another excellent computer program for the electric powered flight enthusiast is offered by Paul Ogushwitz of USR&D. The program is known as "Aero Comp." It is capable of operating on IBM compatible computers using DOS systems of 2.0 or later. I have had no personal experience with this particular program, but have heard many good things about it from fellow electric modelers.

A good digital readout scale is essential for any electric flyer because weight is so much more critical for successful electric performance. The author's Pelouze scale costs about $70 at office supply stores, and can read in grams or ounces up to 80 ounces maximum.

PHOTO: BOB ABERLE

Computer programs can also assist the modeler when it comes to selecting special props for certain scale applications, e.g., the three- and four-bladed variety. Trying to make these selections experimentally can prove to be a big chore, but not for a computer. Equally important is multi-motor operation (twins, three motors, four motors, etc.) and the possibility of one motor driving two props via a belt drive or the converse with two motors connected to drive a single prop at higher thrust levels. All of these seeming complicated variations can best be handled by a computer rather than the classic modeling experimenter.

Many modelers, unfortunately, are still quite intimidated by computers while others who must be content with fixed incomes can't justify the cost of owning their own computer system. For these modelers I might suggest that some enterprising person establish a "call-in" service where, for a small fee, the modeler supplies his electric powered aircraft parameters and the service supplies the recommended choices for a motor, gear drive, prop, and battery pack combination. This might even be set up on a "900" line telephone service.

The final suggestion to help in the electric motor, battery, prop selection process is to deal with a total system supplier, a person who specializes in selling all the necessary components in the electric power system. There is never a substitute for experience.

suitable aircraft

In its early stages of development, electric power aircraft took on two usual forms, sport trainers and powered gliders. The trainers have gotten a lot more sophisticated these days, sporting such things as semi-symmetrical airfoils and aileron flight controls. Electric powered sailplanes have progressed from basic sport/fun flying to new duration oriented competition aircraft. Electric powered flight, however, has gone considerably further than just trainers and sailplanes!

Pylon racing has been developing in Europe using the small size Speed 400 ferrite motors with the total model weight of only 16 ounces. These models are easy to build, inexpensive, and fun to fly. I suspect we will eventually see electric powered "Quickie 500" racers. The beauty of electric power in a racing event is that all motors can be "started" at the same instant every time.

Pattern or aerobatic models are beginning to appear, utilizing the larger size electric motors (e.g., the 25, 40 and 60s). Since battery weight prohibits a full 10-minute flight at maximum power, electric pattern is now being flown in two flight segments. You do half your maneuvers on the first flight, recharge and conclude your maneuvers on the second flight. Generally speaking, whatever maneuver you can do with "wet" power can be duplicated with electric.

The Impulse represents what electric power sailplanes have become. They are no longer the plodding slow "butterflies" of old. Now, like the Impulse they are fast, with amazing thermal performance.

PHOTO: BOB ABERLE

Scale and stand-off-scale models with electric power, are gaining in popularity. Special paints, striping tapes, and decals need not be protected because fuel is not used. Models with short noses (or long tail moments) can be more easily balanced by positioning the heavy battery packs. Best of all is the practicality of multi-motor flying (two engines, four engines, etc.). Each motor runs from the one battery pack. The motors start at the same time and they gradually lose power in unison. There is no chance of one motor stopping on an electric twin. Throttling multi-motors can also be handled by a

Electric powered pylon racres are now being developed like the EMF which Tom Hunt holds. This design by author Bob Aberle uses an direct drive 05 motor, and was published in Flying Models magazine.

PHOTO: BOB ABERLE

As understanding of propulsion systems develops, the size of models matters less. Tony Fiore's beautiful P-51 Mustang is an impressive 1/5 scale. It was modified from a Pica kit originally designed for "wet" power. Tony used a direct drive Astro Flight cobalt 90 with forty 1400 mAh cells, and a 16–12 prop

PHOTO: BOB ABERLE

single controller.

Noted cartoonist/author, Don Bousquet, and his son, Nate, have been flying a very large twin boom electric powered pusher model, powered by an Astro Flight geared cobalt 40. In the nose of this model is a Canon UC-1 8mm camcorder. Total model wing area is 1400 square inches with a gross weight of 13.75 pounds, yet it is as stable as a trainer. The resulting videos taken from the sky have been absolutely spectacular.

Art Thoms has a large scale model of the Boeing Model 314 PAA Clipper which flies off the water, as did the full size original. Electric powered rise-off-water models are popular because there is less chance of a stalled motor preventing you from retrieving your model. Quiet operation and the lack of fuel, which can pollute the water, make this type of flying very practical at local park facilities, which have small lakes or ponds.

Though not the size of a modern Pattern aircraft, Aerocraft's Apache is designed for aerobatic flying. At 44-inch span, it has 340 square inches of wing and flies with an 05 geared motor.

PHOTO: AEROCRAFT MODELS

Electric helicopters have been a reality for some time. Just think of the possibility of learning to hover your helicopter while in the confines of your own backyard and without creating any noise that would ever bother a neighbor.

Freeflight electric powered flying is another popular sport. These models fly on their own without the aid of a radio control system. Some very small direct and gear drive motors are now available which can power miniature scale models that weigh no more than a couple of ounces.

Dick Miller writes an electric flight column for *Flying Models* magazine, and has been specializing in both these small electric freeflights and some equally small electric R/C models.

I recently came upon a tiny (½ ounce)

Don Bousquet (L), with his son Nate (R), proves what electric models can do with his original twin boom R/C aircraft . It carries a Canon UC-1 8mm camcorder in its nose for inflight pictures. The combination of a single Astro 40 geared motor and twenty-one 1700 mAh cells provide the thrust for the 13-3/4 pound model.

PHOTO: BOB ABERLE

Another multi-engine wonder is Dr. Keith Shaw's flying wing. It is powered by four Leisure 05 geared ferrite motors, and also has operating retracts. Keith has been in the forefront of electric powered flight, and has created a wide variety of unique and impressive R/C aircraft.

PHOTO: BOB ABERLE

solid state electronic timer that can be used to shut down an electric motor after a pre-set timed period. This is perfect for test flying since you can initially set the motor to shut down after only a few seconds and as you continue to trim and refine the flight characteristics, the motor run duration can be increased.

The switch is known as the Snappy Free Flight Timer and can be obtained from Gordon Tarling. These switches can handle from four to seven cells at up to 7 amps, and can time up to approximately 50 seconds.

Controline flying is still another form of modeling that can make good use of electric power. Think of the possibility of flying in populated areas without causing any noise. You could even try the tethered type of flying where the control wires are attached to a centrally located pylon then a second set of wires are run to a point outside the circular path of the model. The advantage of that type of flying is that it avoids the dizziness that comes from flying in circles.

We are now seeing a new direction in electric powered flight devoted to ducted fans and, in some cases, shrouded prop propulsion systems. This is not a lark by any means, as I have seen several ducted fan electrics that were absolutely spectacular.

Dr. Keith Shaw, the famous electric flyer from Michigan, has a scale model of the German Horten flying wing powered by two electricducted fans. Another famous flyer, Tom Hunt, has an electric ducted fan powered replica of the WW

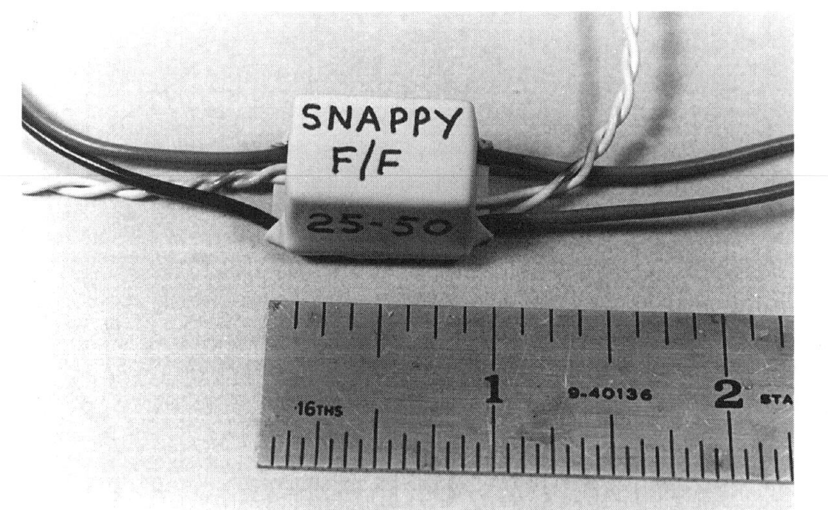

The Snappy Freeflight Timer was designed by Gordon Tarling. This 1/2 ounce device can time electric motors from five, up to 50 seconds. It can handle from four to seven cell batteries and a maximum current of up to seven amps.

PHOTO: BOB ABERLE

Perhaps the one area of radio control that has and will benefit the most from electric power is a multi-engine model, like Ken Stinson's C-130 Hercules above. It uses four Astro 035 geared motors fed by a 20-cell 1400 mAh battery. Total weight is 8-1/2 pounds.

PHOTO: BOB ABERLE

These days, nothing escapes the scope of electric power, including helicopters. This Hughes 500 is an electric powered chopper from Hobby Lobby. It sells for under $200 and uses a Graupner Speed 600 motor fed by a 7-cell 1700 mAh battery.

PHOTO: BOB ABERLE

II German V-1 Buzz Bomb and it is very realistic! HiLine Ltd. offers a tiny ducted fan unit for use with freeflight scale models weighing only a couple of ounces. Only time will tell what new ducted fan applications will appear.

Indoor electric powered aircraft are gaining popularity in cold climates where it normally isn't practical to fly from December until late March. Suitable "flying sites" have been found in school auditoriums, gymnasiums, sports arenas, and old aircraft hangars that have been declared surplus property by the government. Both freeflight and R/C models are being flown indoors using electric power.

Several single channel ultra light weight radio systems have surfaced in the past year. It is now possible to have an electric powered single control R/C model with an all-up weight of under 2 ounces that can fly for considerable periods of time. There is a lot of fun to be had in this facet of the electric powered flight hobby.

Finally we have those wonderful novelty type models that are also benefitting

Free Flight Nostalgia designs have become popular choices for electric conversion. Bob Aberle's Infant Sportster was designed by Bill Winter in 1949 for glow power. Bob installed an IMP-30 motor with three 700 cells, along with R/C rudder, elevator, and motor control. Total weight of the model is 14 ounces.

PHOTO: BOB ABERLE

from electric power. Altech Marketing's ParaPlane Sport, which is based on a very similar, real aircraft, is a good example of this. An electric driven R/C controlled power pod is suspended from a large parachute called a para-foil. Steering control is achieved by rocking the parachute support bar. Elevation is controlled by a motor speed controller.

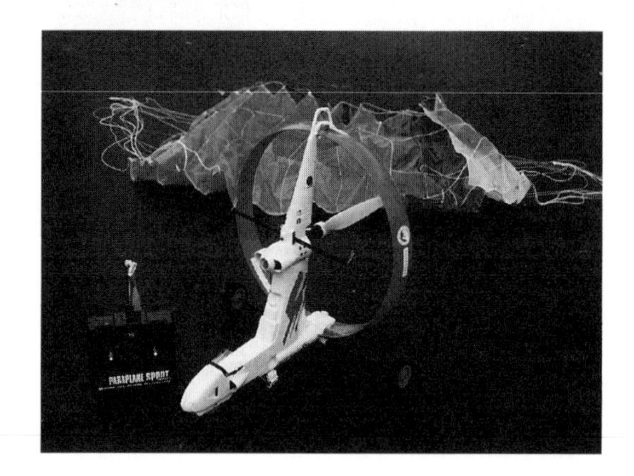

Altech Marketing sells this electric R/C sport scale model called the Paraplane Sport. Like the real aircraft, the shrouded prop pushes the R/C Paraplane Sport forward while the special parafoil provides the lift.

PHOTO: BOB ABERLE

Equally popular are such items as electric powered hang gliders that look very much like ultra-lights and radio controlled helium filled blimps or dirigibles, which are powered by two shrouded props turned by a pair of electric motors. Steering control of these blimps is achieved by forward and reverse control on the motors, either in unison, or by having one go in the forward direction, while the other is reversed.

aircraft selection

The Skyvolt, designed specifically for electric power by Bob Kopski, is an advanced trainer that was first published years ago in Model Aviation. It is now kitted by Today's Hobbies and uses up to 15-size motors. It has 400 square inches of wing.

PHOTO: BOB KOPSKI

Slightly larger than the Skyvolt, Bob Kopski's Revolt at the right has 600 square inches of wing and will take from 05- up to 25-size motors.

PHOTO: BOB KOPSKI

Now that you have seen the possibilities for electric powered flight, I think it is important to give some recommendations for a good starting point. If you are entering R/C model aviation and electric power all at the same time, then you must certainly consider a trainer type model. Even if you already have some gas powered model aircraft experience you would be wise, when venturing the first time into electric power, to try a trainer type model just to gain some familiarity.

There are some good kits available for that purpose. One of the most popular is the Skyvolt, designed by Bob Kopski,

A trainer, or a leisurely sport plane, Aerocraft's new Begin-Aire reflects a very popular application of electric power, sailplanes. At 25 ounces the model flies on a Speed 400 direct drive motor, and a 5-cell battery.

PHOTO: COURTESY AEROCRAFT

Electric Columnist for *Model Aviation*. This design was first published in the January 1990 issue of that magazine (for which full size plans are available). It was later offered as a complete kit by Today's Hobbies Inc. It has a 400-square inch wing, can be flown with a geared 05 or 15, and weighs approximately 50 ounces. It also has aileron control which can be very effective even on a trainer.

A larger, more updated version of this model, also designed by Bob Kopski and known as the Revolt, was recently published in the November 1994 Model Aviation. By the time you read this it might be available in kit form as well. The Re-

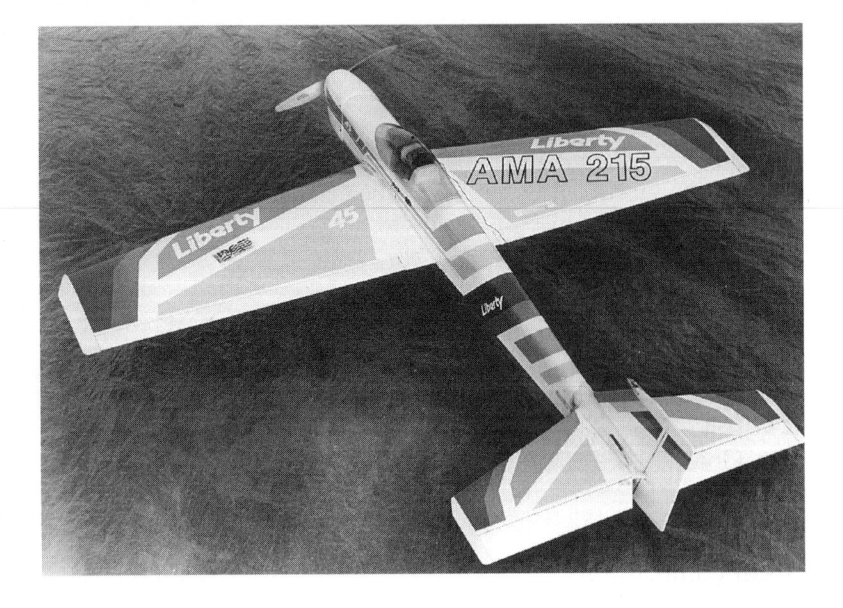

ARF's are generally considered heavier than built-up aircraft. Bob took the Liberty 45, originally a glow engine powered ARF (almost ready to fly), and successfully converted it to electric. It was designed for sport pattern flying and used a .45 glow engine. A MODELAIR-TECH H-1000 belt drive with a single Graupner Speed 700 motor is now installed.

PHOTO: BOB ABERLE

volt has a 600-square inch wing with sufficient dihedral to make it fly well on strictly rudder and elevator control (ailerons not required). It was also designed in such a way as to accept all motors from 05 up to as large as a 40 (if you believe!). It also makes a perfect test bed aircraft to try out all of your new "electric" gadgets.

This shows the original .45 glow engine that powered the Liberty 45. Note the cowl cutouts for cylinder head, muffler, needle valve and carburetor.

PHOTO: BOB ABERLE

Thanks to the clean installation of the H-1000 belt drive, no additional cutouts were required in the original cowl though the cutouts did serve to allow cooling air to circulate in and out of the motor compartment.

PHOTO: BOB ABERLE

As previously mentioned in this book, the Tower Hobbies Primary Trainer Electric (PT-Electric) is another excellent choice when used with a gear driven motor. Tower also offers a very attractive Electri-Cub which is a 480-square inch stand-off-scale Piper Cub for 05 electric power.

Electric powered sailplanes also make wonderful trainers because they fly at slower, more gentle speeds. All the basic turning maneuvers and landings can be mastered with this type aircraft. Models that come to mind are the Airtronics Eclipse, the Aerocraft Begin-Aire, and the Goldberg Electra. Later, as you gain flying experience, these same electric sailplanes are quite capable of flying in local competitions.

Also available, for the more experienced builder, are published plans of electric powered models. Most of the popular model aircraft publications offer plans catalogs which can be most helpful when searching for new construction projects. Plans generally run between $10 to $15 per set. The material must all be purchased separately by the builder. If you are lucky enough to obtain a back issue of the publication containing the entire construction article, you will essentially have your own instruction booklet.

Several trainer/sport type electric designs that come to mind are Bill Winter's Lightning Bug (June 1991 *Flying Mod-*

converting the Liberty

To mount the H-1000 belt drive, two ply plates sandwich the drive's brackets. Each plate has a rail on it.

The belt drive is screwed to the rails, and 90° angle brackets added to the rear of the ply plates.

The angle brackets secure the mount assembly to the firewall. The motor has been removed for clarity.

els) and Bill's Sandpiper (March 1993 *Flying Models*). For those interested in an extremely easy to build old timer model, which also makes an excellent electric trainer, try the Schmaedig Stick (January 1992 *Flying Models*). Another excellent choice is the late Ken Willard's Electromaster (November 1989 *RC Modeler*).

Keep in mind that these recommendations are for basic and/or training type model aircraft. As you get more involved it will become obvious that many, many kits and published plans are available for all kinds of electric powered aircraft, from beginner to expert.

Another type of model worth mentioning is generally referred to as the "ARF" (Almost-Ready-To-Fly). These prefabricated models are usually supplied already covered and require a minimum of assembly time. The typical assembly involves joining the wing halves and installing the motor and the radio system! Unfortunately for the electric flyer, there has not been much effort to date on the part of manufacturers to supply ARFs expressly for electric power.

Wanting to prove a point, I took on the job recently of converting a popular .45 glow engine powered ARF sport/pattern model into electric power. The project took less than a day and turned out, by my standards, to be very successful. A description of what I did might prove helpful for other conversion projects that you might want to consider.

The model chosen was the Liberty 45, which was offered several years ago by one of the major national hobby distributors. The exact model is no longer available, but several similar ones are. The Liberty is a 630-square inch low wing tail dragger configuration. All-up weight with a .45 glow engine/muffler was 5.24 pounds, with a relatively light wing loading of 19.1 ounces/square foot. Not wanting to take the time to do the experimenting myself, I decided to try a new hobby oriented company that is dedicated to electric powered flight known as Modelair-Tech.

Tom Hunt of this firm has developed and produced a very interesting belt drive assembly which he calls his H-1000. The 1000 refers to the fact that this drive can handle electric motors up to 1000 watts rating. It can use one single 1000-watt motor or it can be configured to take two separate motors (up to 500 watts each) which drive a single prop shaft. Besides selling the belt drive, Modelair-Tech will provide you with enough information to convert many of the current gas powered kits and ARF's currently on the market to electric power.

When Tom Hunt was given the parameters for my Liberty 45, he came up with an H-1000 belt drive with a 2.57 to 1 reduction ratio to be powered by a Graupner Speed 700 ferrite motor. The combined cost of the belt drive and Speed 700 motor is inexpensive! Tom went on to suggest a 12–8 prop so that ground clearance would still make take-offs and landings very practical.

For a battery pack it was suggested that I initially try a 14-cell SR 1100 pack as a means of verifying the potential performance. The gas engine, muffler, fuel tank, motor mount and R/C throttle servo were removed. A radial mount adapter was fashioned from ⅛-inch plywood to accept the beam mounting holes on the H-1000. This adapter was then bolted to the ARF's plywood firewall. Not knowing exactly where the battery pack would be located to achieve roughly the same c.g. location, I chose to place the radio receiver and battery pack up inside the clear canopy. For a motor controller I selected the new AI Robotics FX-35D, which was set up with the in-flight radio battery charging feature as described in **Chapter 6**.

Total weight of the converted Liberty 45, with the 14-cell 1100 battery turned out to be 6.4 pounds with a wing loading of 23.6 ounces/square feet. Flight performance was good. The Liberty lifted off our grass runway after about a 75-foot run and flew at full throttle for about 2½ minutes. After that first flight, I felt that the plane could use more power, needed more duration, and it could easily handle more weight.

The final step in this conversion process was to substitute a 16-cell pack consisting of SCRC cells rated at 1700 mAh. Going to 16 cells also warranted going from a belt reduction ratio of 2.57:1 to 3.0:1. The added weight of this pack brought the Liberty 45 to a total of 7 pounds. It was also necessary to move that pack aft somewhat to reduce what looked like a definite nose heavy tendency. The wing loading at this new weight worked out to 25.6 ounces/square feet. With this larger battery pack, the Liberty was off the ground in about 35 feet and now has the capability of pulling vertical maneuvers for at least half of the charge. Total duration at full throttle is about four minutes; with some throttling back, five minutes is easily attainable.

To me this conversion process was a total success. It should open the door for many modelers who might have a favorite gas powered aircraft, but wish it was electric powered. Using this belt drive it is possible to convert gas powered ARFs, regular kits and even published designs all to electric.

flying

After learning as much as you can about electric power, selecting your first model aircraft, building it, and installing the equipment, you are now ready to fly. As I stated at the very beginning, you don't have to be an electrical engineer to enjoy electric powered flight. In fact, you could make your very first R/C model aircraft flight using electric power. There is no prerequisite that says you have to fly "wet" power first.

However, the most important bit of advice I can give you at this point is to seek help from an experienced modeler. Learning to fly R/C models works best as a team effort where the team consists of an instructor pilot, and you as the student.

The best starting point is to find a local R/C flying field which will usually be operated by a model airplane club. Your local hobby shop is a good place to ask about flying fields and clubs.

With a simple flip of the switch the instructor pilot on the left can take over instant command from the student pilot on the right. The instructor holds the master transmitter, while the student holds the slave transmitter. Identical brand transmitters must be used for this function.

PHOTO: BOB ABERLE

The next step is to join the Academy of Model Aeronautics. This is the national organization of model aviation, and represents any type of model that flies—radio control, freeflight, or control line—and in any form—helicopters, gliders, electric power, ducted fans, etc. Membership in the AMA presently costs $42.00 annually (1995) and the address is 5151 East Memorial Drive, Muncie, IN 47302 (telephone: 1-800-FLY AMA). For this fee you will receive an excellent monthly magazine called Model Aviation, plus a high value supplemental liability insurance policy, and a rule book (on request) which contains all the details of the various organized competitions that are coordinated by the AMA. As an AMA member you are entitled to purchase an excellent book, titled, *Basic Flight Training*, by Phillip Balcomb, for a special price of $5.00. Another benefit of AMA membership is that you can call into their headquarters and the staff will help you locate a local club, as well as a flying field. This is done through a computer system that is keyed to local zip codes. You supply your own zip code and the computer can tie you in to the clubs and flying fields in your areas. You may even be able to find local clubs that specialize in electric powered flight.

Another way of locating local clubs, flying fields, and possibly instructor pilots is by using a national computer online service. There is an excellent and very popular network known as "Modelnet" which operates on the CompuServe system. You might consider participating in this network, as well as joining the AMA.

Once you have located a local club, it would be wise to make an initial visit to their flying field with the idea of learning the safety rules and procedures as each club will have their own specific rules. Very important on the list of items to learn initially is radio frequency control. If you turn on your R/C transmitter and it is operating on the exact same channel as someone who was currently flying, it is likely you could interfere with that person and possibly crash his plane. Frequency control is a very serious matter and should be your first concern even before you attempt to assemble your model for its first flight.

Another important item for that first time out is to have a club official or an experienced modeler make a safety inspection of your model. The usual procedure is to pull on your control surface hinges to make sure they are anchored properly. Wing attachments are checked to make sure your wing doesn't part company in flight. Even the motor mount will be checked. The reason for this is to verify the structural integrity of your model aircraft with a view towards overall safety.

One of the important things you will be shown prior to first flight is how to check the center of gravity (c.g.) or balance point of your model. Every model must be balanced according to its designer's instructions before the first flight is ever attempted. You might want to bring the plans for your model with you on that first visit to the flying field. Electric powered model aircraft are usually the easiest to balance properly because the mass of the battery pack can easily be shifted around inside the fuselage. If the model is nose heavy, move the pack aft; if the model is tail heavy, move the battery pack forward.

In addition to obtaining the assistance of a local instructor/pilot, there are some helpful products you may want to consider when learning to fly an R/C model for the first time. Several excellent computer driven simulation systems exist on the hobby market. One in particular that comes to mind is offered by Dave Brown Products Inc. It includes a control box that resembles a basic R/C transmitter with actual control sticks. Moving these sticks allows you to "fly" the model airplane image on the monitor just as you would at the field. You use this simulator much as you would a video game program.

Although the majority of instructor/pilots can only stand next to their students ready to take the transmitter if the student gets into trouble, there is another option. There is a training system which involves the use of a special cable that can hook two transmitters together, if they are the same brand. If both you and your instructor have the same brand of R/C system, both transmitters can be coupled together with this training cable. This allows each of you to physically hold a transmitter and finger the control sticks.

The instructor's transmitter (the master transmitter) must be on the same frequency as the receiver in the model. The frequency doesn't matter on the student's transmitter (slave transmitter). There also will be a master/slave switch on the mas-

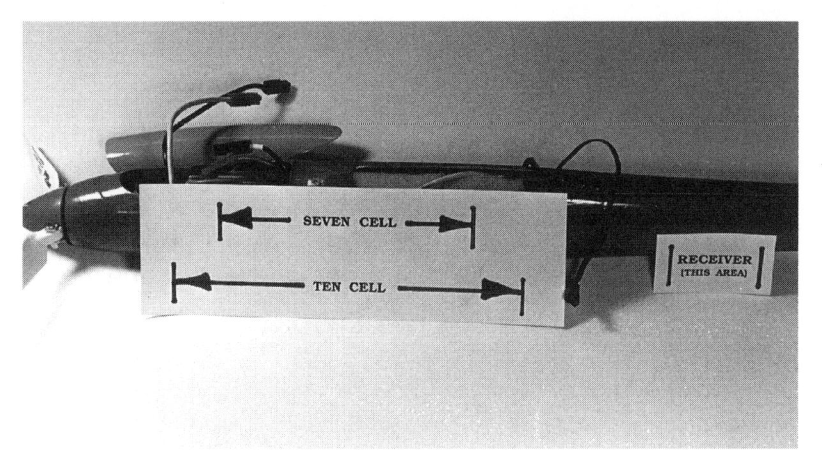

Obtaining the proper c.g. (center of gravity) on electric models is relatively easy. Simply shift the mass of the motor battery forward or aft as needed. The photo shows the location for a 7-cell or 10-cell battery.

PHOTO: BOB ABERLE

ter transmitter. When the instructor engages it, the student is in control of the model while the instructor monitors the flight. If the student gets into trouble, the instructor releases the master switch and immediately takes over the flight. There's no delay in handing the transmitter back to the instructor.

The length of time it will take to comfortably solo your model (take off, fly around, and land where you want it to land) depends on the individual. Younger modelers tend to learn faster (maybe because they have more nerve!). One of the hardest things to master when flying a radio controlled model airplane is to remember that when the model is coming at you, the perception of turning or steering controls are reversed (left is right and right is left!). If you were flying a full size plane and sitting in the cockpit, that would not be the case.

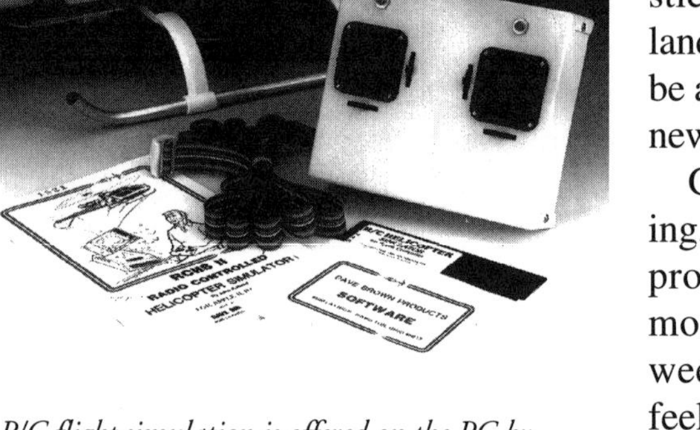

R/C flight simulation is offered on the PC by Dave Brown's Flight Simulator. It's a "painless" way to commit pilot errors, and learn.

PHOTO: COURTESY DAVE BROWN PRODUCTS

It has been my experience that full scale pilots learning R/C flight often have trouble with this "turning phenomena." A good instructor/pilot should be able to help you with the mechanics of steering a model while it is heading towards you. After a while it will come naturally and you won't have to think about it.

I might add that there are certain advantages to learning to fly R/C while using electric power. When flying a gas or wet powered model it is always possible to have a "flame-out" (engine stopping for no apparent reason in flight). This can happen when you run out of fuel or simply have the engine tuned improperly (poor idle adjustment on the carb!).

With electric power you will gradually see the model slowing down in flight as the battery charge is depleted. This should make it obvious as to when you should be getting ready for your landing. Another advantage: you will never lose an engine abruptly, as you might with gas power. Therefore, it is not likely that you will have to make any "dead-stick" (engine-out) landings, which can be a real plus for the new R/C pilot.

Generally speaking, it will take approximately two months or so of weekend flying to feel comfortable at the controls without help. I had one fellow in his mid-twenties who I taught to fly in two weeks of intensive lunch hour training sessions. In about eight sessions he was on his own. Concentrated training sessions tend to accelerate the progression, speeding things up. Spreading out the sessions tend to slow things down. Unfortunately, bad weather and scheduling problems can hinder the process at times. The most important thing to remember is that it is a hobby and it should be enjoyable.

getting to know electrics more

After you have succeeded with electric powered flight and are comfortable flying solo, you will want to enhance your knowledge of this new skill. As you have already learned, there are many types of model aircraft that can be flown with electric power, many of which require special skills that you will achieve once you gain experience.

One of the most informative and oldest electric flight publications is Electric Flight U.K., the official newsletter of the British Electric Flight Association.

My first suggestion is that you regularly read the various electric columns offered in the model press. Here is a list of these columnists and their publications: Bob Kopski (*Model Aviation*), Larry Sribnick, Dick Gibbs and Dick Miller (*Flying Models*), Ted Davey *(RC Modeler)*, Dr. Keith Shaw, Mitch Poling and Dave Baron (*Model Airplane News*), Dr. John Mountjoy (*R/C Reports*) and Roger Jaffe (*Model Builder*). In the process of scanning these publications, you will begin to note construction articles of electric powered aircraft where full size plans are offered and also product reviews of new kits and equipment are presented.

Although at this time there is no magazine dedicated to electric power in the United States, there are two very excellent publications that are published in Europe. The first is known as *Quiet Flight International* and the other, which started with the November/December 1994 issue, is called *Electric Flight In-*

In recent times, two new publications, Electric Flight International and Quiet Flight International have emerged, which deal primarily with R/C electric flight models.

PHOTO: BOB ABERLE

Probably the preeminent electric flight gathering in the world, the KRC Electric Fly, held in Quakertown, PA, brings hundreds of R/C electric flight enthusiasts together to share ideas and inspiration.

PHOTO: BOB ABERLE

One of the more popular events at the KRC Electric Fly is the "All Up-Last Down" event. Here, 25 models at the 1994 KRC meet are launched in this competition.

PHOTO: BOB ABERLE

ternational.

Another good source of electric power technical information are the various electric club newsletters which are circulated nationally. There are many clubs chartered expressly for electric powered flight. You can learn where to obtain these newsletters by reading the magazine columns just mentioned or by word of mouth from various clubs. Most of these clubs will offer subscriptions for their newsletters at modest costs so that you may receive them through the mail on a regular basis.

Without a doubt, the very best newsletter (although it actually looks like a magazine) is the *Electric Flight U.K.*, the official publication of the British Electric Flight Association (BEFA). The editor of this journal is Dave Durnford and he does an excellent job with this publication. It is definitely worth reading. You can make inquiries to Dave concerning the availability of this newsletter by writing to him at the address listed in the appendix.

One of the best ways to exchange technical knowledge concerning electric powered flight is to attend the annual KRC Electric Fly which is sponsored by the Keystone R/C Club and held every September in Quakertown, PA. This is, without a doubt, the largest electric power only gathering in the world. Each year hundreds of electric flyers get together for a weekend of fun.

Although several competitions are held, the main theme is to just go out there and show off your latest electric powered creation. A detailed article on all aspects of this KRC Electric Fly appeared in the February 1995 issue of *Flying Models* magazine. I suggest you try to obtain a back issue. Make your reservations early if you plan to attend since this is a very popular annual event.

At this same KRC Electric Fly in Quakertown, SR Batteries Inc. sponsored their first Symposium on Electric Powered Flight at the September 1994 event. Seven guest speakers presented technical papers on various subjects of electric flight on Friday, the day before the actual KRC Electric Fly begins. This symposium will now become a regular feature on that same weekend. Reservations to attend this symposium must be made

Dr. Keith Shaw (L) and Steve Anthony (R) field questions from the audience at the 1994 SR Batteries Symposium held in conjunction with the 1994 KRC Electric Fly.

PHOTO: BOB ABERLE

Both the AMA and SAM (Society of Antique Mode-lers) actively incorporate electric flight enthusiasts. AMA, which is the national model organization, also provides insurance coverage.

PHOTO: BOB ABERLE

well in advance directly with Larry Sribnick of SR Batteries Inc.

To help expand your knowledge and participation in electric flight it would be helpful (and fun) to attend the various contests or competitions that are held each year throughout the country. Most of these events will be advertised in *Model Aviation*, the magazine of the AMA, in their Competition Newsletter section. Of course, the various club newsletters which are networked throughout the country, also do a considerable amount of advance advertising for electric contests.

The popular events generally flown include electric powered sailplane in Class A (up to 7-cell batteries and 45-second motor run), and Class B (more than 7 cells and a 30-second motor run). Both are AMA events, and the detailed rules are published in the AMA Rule Book. AMA will supply this book for a nominal fee.

Also very popular are the electric Old Timer events. The AMA has several of these which involves the flying of models designed prior to the start of WW II. The Society of Antique Modelers (SAM) has included two other electric powered Old Timer events which have likewise become popular. You can join the SAM organization for an annual $15.00 fee by writing to: Mr. Larry Clark, Treasurer, PO Box 528, Lucerne Valley, CA 92356.

Mention should also be made of a truly fun type event which has become known as the "All Up–Last Down." This is a great spectator event, since all the participating models are launched simultaneously. Trusting that enough different R/C channels are available, it is common to see as many as 15 to 20 models all launched at the same time. The rules are easy as the last person down is the winner and flights of an hour or more are common. Battery management is the crucial part of this competition. Most of the Electric Fun Fly gatherings will have at least one of these events each day.

future of electric flight

In my personal opinion, two major items are necessary for the future growth of electric powered flight. The first is

Larry Sribnick, founder of N.E.A.C., the National Electric Aircraft Council, takes the podium at the 1994 SR Batteries Symposium to welcome those attending. It was held in conjunction with the 1994 KRC Electric Fun Fly.
PHOTO: BOB ABERLE

the formation of a properly organized and staffed national organization to represent electric power enthusiasts in the United States. The second, and equally important, is the encouragement of new technology to get the very most out of electric power.

Some years ago, several attempts were made to form a national electric association as a special interest group within the overall hobby of model aviation. None of these organizations succeeded in the long term, for many reasons. Instead of looking back, the time has come to look to the future and be reminded that when hundreds of electric power enthusiasts can gather at the big KRC Electric Fly every year, there is definitely enough interest to establish a solid foundation for a national electric organization.

The timing of this book was just perfect because it will allow me to introduce our newly formed national electric association. This information was learned just as I completed the previous paragraph. It is called the "National Electric Aircraft Council" or NEAC. In January 1995, the AMA Executive Council approved the NEAC as the official "SIG" (Special Interest Group) to represent electric flyers throughout the United States. The founding Chairperson of NEAC is Larry Sribnick. Any modelers wishing to learn more about the organization and how to join, can write to Larry at SR Batteries.

I hope that NEAC will encourage the growth of electric powered flight by standardizing such items as: motor sizing (that will be recognized and understood by all), battery capacity ratings, safety requirements, rules for competitions, and a

N.E.A.C.

structure that can sponsor regional and national electric powered aircraft contests.

NEAC will also interface with counterpart organizations throughout the world. As previously mentioned, the British Electric Flight Association (BEFA) is a model of excellence in organization and probably the reason why British electro-modelers already enjoy two dedicated electric flight magazines. We now look forward to such possibilities catching on in the United States.

A national electric powered flight newsletter (sponsored by NEAC) will help the second wish which is for technology growth. Over the long term I'm sure we'll see things as more efficient motors, and higher capacity, lower weight batteries.

But in the short term there is much we can do in the name of improvement. Using our available electric energy properly is something we can all learn better and we can do it now. Selecting the proper gear or belt motor drive for optimum performance is a key factor. Innovative motor drive systems that permit the stacking of several motors on a single shaft for increased output or a single motor driving several props along with the expanded use of multi-motor operation can all contribute to the growth of electric flight. Let's not forget the application of electric powered ducted fans for various scale applications.

Increased participation in electric power will improve the market potential. As sales increase, more manufacturers will enter the market, competition will increase and prices will begin to go down. My feeling is that more and more model aircraft, of varied types up to as large as quarter scale, will be flown regularly with quiet, clean electric power.

I also look forward to "ready to run," complete electric power systems. If a person can purchase a radio control system today and simply plug all the components together in a matter of minutes, why can't the same be provided for electric power? In other words have the motor, prop drive, speed controller and battery sold in a single package with all the connectors already installed. No soldering would be necessary. If you agree with any of these ideas I urge you to write to the manufacturers listed at the end of this book and share your thoughts with them.

Please lend your support for the new National Electric Aircraft Council by both joining and, most important, by actively participating.

Good Electric Flying to All.

Motors

Astro Flight Inc.
13311 Beach Ave.
Marina Del Ray, CA 90292

Aveox Inc.
P.O. Box 1287
Agoura Hills, CA 91376-1287
(brushless DC motors and controllers)

Cermack Electronics & Model Supplies Co.
107 Edward Ave.
Fullerton, CA 92633

Flightec Inc.
21 Juniper Way
Hamilton, NJ 08619
(brushless DC motor and controller)

HiLine Ltd.
P.O. Box 11558
Goldsboro, NC 27532
(very small ferrite motors)

Hobby Lobby International
5614 Franklin Pike Circle
Brentwood, TN 37027
(source for Graupner motors of all types)

Kenway Micro Flight
P.O. Box 889
Hackettstown, NJ 07840
(tiny ferrite motors and gear drives)

Kress Jets
500 Ulster Landing Road
Saugerties, NY 12477
(custom motors, special applications)

Leisure Electronics
2950 Airway Bldg., A-4
Costa Mesa, CA 92626

MaxCim Motors
57 Hawthorne Drive
Orchard Park, NY 14127-1958
(brushless DC motors, controllers)

MODELAIR-TECH
P.O. Box 12033
Hauppauge, NY 11788-0818
(ferrite motors)

Model Electronics Inc.
6500 6th Ave.
Seattle, WA 98117
(ferrite motors)

New Creations R/C
P.O. Box 496
Willis, TX 77378

Pica-Robbe Inc.
2655 N.E. 188th Street
Miami, FL 33180

SR Batteries Inc.
P.O. Box 287
Bellport, NY 11713
(neodym motors)

Tower Hobbies Inc.
P.O. Box 9078
Champaign, IL 61826-9078

VL Products Inc.
7871 Alabama Ave.
Canoga Park, CA 91304
(miniature electric motors and gear boxes)

Props, Adapters, Spinners

There are also many spinners available for use with both fixed and folding blade props. The following manufacturers/distributors can supply props, adapters and spinners for electric power use.

HiLine Ltd.
P.O. Box 11558
Goldsboro, NC 27532
(electric ducted fans)

Hobby Lobby International
5614 Franklin Pike Circle
Brentwood, TN 37027
(Graupner Scimitar folding props, Freudenthaler and Aeronaut folding props)

Kress Jets
500 Ulster Landing Road
Saugerties, NY 12477
(electric ducted fans)

Landing Products
P.O. Box 938
Knights Landing, CA 95645
(APC plastic and composite material props)

Slegers International
Route 15
Wharton, NJ 07885
(Freudenthaler carbon blade folding props)

SonicTronics Inc.
7865 Mill Road
Elkins Park, PA 19117
(folding props)

SR Batteries Inc.
P.O. Box 287
Bellport, NY 11713

Windsor Propeller Co.
3219 Monier Circle
Rancho Cordova, CA 95742
(Master Airscrew electric wood props and folding prop assemblies)

Batteries & Chargers

ACE R/C Inc.
116 W. 19th Street
Higginsville, MO 64037

Astro Flight Inc.
13311 Beach Ave.
Marina Del Ray, CA 90292

B&T R/C Products Inc.
508 Lake Winds Trail
Rougemont, NC 27572

Cermack Electronics & Model Supplies Co.
107 Edward Ave.
Fullerton, CA 92633

D.A.D. Inc.
168 Main St.
Chadron, NE 69337

HiLine Ltd.
P.O. Box 11558
Goldsboro, NC 27532

Hitec RCD Inc.
10729 Wheatlands Ave., #C
Santee, CA 92071

Hobby Lobby International
5614 Franklin Pike Circle
Brentwood, TN 37027

Kress Jets
500 Ulster Landing Road
Saugerties, NY 12477

Pica-Robbe Inc.
2655 N.E. 188th Street
Miami, FL 33180

SR Batteries Inc.
P.O. Box 287
Bellport, NY 11713
(specializes in custom battery packs)

Tower Hobbies Inc.
P.O. Box 9078
Champaign, IL 61826-9078

TNR Technical Inc.
279 Douglas Ave.
Altamonte Springs, FL 32714

E. F. Yost & Co.
2211 D Parview Road
Middleton, WI 53562

Controllers

ACE R/C Inc.
116 W. 19th Street
Higginsville, MO 64037

AI-Robotics Inc.
1440 Third St.
Suite 4
Riverside, CA 92507

Airtronics Inc.
11 Autry
Irvine, CA 92718

Astro Flight Inc.
13311 Beach Ave.
Marina Del Ray, CA 90292

Aveox Inc.
P.O. Box 1287
Agoura Hills, CA 91376-1287

Cermack Electronics & Model Supplies Co.
107 Edward Ave.
Fullerton, CA 92633

Flightec Inc.
21 Juniper Way
Hamilton, NJ 08619

High Sky R/C Accessories
322 Maple Lane
Nappanee, IN 46550

Hitec RCD Inc.
10729 Wheatlands Ave., #C
Santee, CA 92071

Jomar Products Inc.
Div. of Electronic Model Systems Inc.
22483 Mission Hills Lane
Yorba Linda, CA 92687

Lofty Pursuits Inc.
2274 Aster Ct.
Rapid City, SD 57702

MaxCim Motors
57 Hawthorne Drive
Orchard Park, NY 14127-1958

New Creations R/C
P.O. Box 496
Willis, TX 77378

Pica-Robbe Inc.
2655 N.E. 188th St.
Miami, FL 33180

Slegers International
Route 15
Wharton, NJ 07885

SR Batteries Inc.
P.O. Box 287
Bellport, NY 11713

Gordon Tarling
87 Cowley Mill Rd.
Uxbridge, Middlesex
Great Britain UB8 2QD

Gear & Belt Reduction Drives

Anthem Metal Products Inc.
42 A Nancy St.
West Babylon, NY 11704

Astro Flight Inc.
13311 Beach Ave.
Marina Del Ray, CA 90292

Kenway Micro Flight
P.O. Box 889
Hackettstown, NJ 07840

Leisure Electronics
2950 Airway Bldg., A-4
Costa Mesa, CA 92626

MODELAIR-TECH
P.O. Box 12033
Hauppauge, NY 11788-0818

Model Electronics Inc.
6500 6th Ave.
Seattle, WA 98117

Pica-Robbe Inc.
2655 N.E. 188th Street
Miami, FL 33180

Windsor Propeller Co., Inc.
3219 Monier Circle
Rancho Cordova, CA 95742

Connectors & Wire

ACE R/C Inc.
116 W. 19th Street
Higginsville, MO 64037

Astro Flight Inc.
13311 Beach Ave.
Marina Del Ray, CA 90292

W.S. Deans & Co., Inc.
7628 Jackson St.
Paramount, CA 90723

Radio Shack Stores Inc.
(local yellow pages)

Sermos R/C Snap Connectors Inc.
Cedar Corners Station
Box 16787
Stamford, CT 06905

SR Batteries Inc.
P.O. Box 287
Bellport, NY 11713

Tower Hobbies Inc.
P.O. Box 9078
Champaign, IL 61826-9078

Radio Systems

Airtronics Inc.
11 Autry
Irvine, CA 92718

Cannon R/C Systems Inc.
3759 N. Green Valley Parkway
Henderson, NV 89014

D.A.D. Inc.
168 Main St.
Chadron, NE 69337

Futaba Corporation of America
4 Studebaker
Irvine, CA 92718

Hitec RCD Inc.
10729 Wheatlands Ave., #C
Santee, CA 92071

JR Remote Control
c/o Horizon Hobby Distributors
3102 Clark Road
Champaign, IL 61821

Polk's Modelcraft Hobbies
346 Bergen Ave.
Jersey City, NJ 07304

Kits & Semi-kits

Aerocraft Inc.
P.O. Box 553
East Northport, NY 11731

Airtronics Inc.
11 Autry
Irvine, CA 92718

Hitec RCD Inc.
10729 Wheatlands Ave., #C
Santee, CA 92071

Hobby Lobby International
5614 Franklin Pike Circle
Brentwood, TN 37027

Kress Jets
500 Ulster Landing Road
Saugerties, NY 12477

Midwest Products
400 South Indiana St.
Hobart, IN 46342

MODELAIR-TECH
P.O. Box 12033
Hauppauge, NY 11788-0818

Model Electronics Inc.
6500 6th Ave.
Seattle, WA 98117

New Creations R/C
P.O. Box 496
Willis, TX 77378

Pica-Robbe Inc.
2655 N.E. 188th Street
Miami, FL 33180

Slegers International
Route 15
Wharton, NJ 07885

Today's Hobbies Inc.
41 E. Lincoln Ae.
Hatfield, PA 19440

Tower Hobbies Inc.
P.O. Box 9078
Champaign, IL 61826-9078

Miscellaneous

Dave Brown Products Inc.
4560 Layhigh Road
Hamilton, OH 45013
(R/C computer simulator, light wheels)

HiLine Ltd.
P.O. Box 11558
Goldsboro, NC 27532
(plans)

Kenway Micro Flight
P.O. Box 889
Hackettstown, NJ 07840
(plans)

Kress Jets
500 Ulster Landing Road
Saugerties, NY 12477
(performance analysis computer program)

MODELAIR-TECH
P.O. Box 12033
Hauppauge, NY 11788-0818
(plan service)

Radio Shack Stores Inc.
(local yellow pages)
(wire, connectors, soldering supplies)

U.S.R.&D. Inc.
P.O. Box 753
Hackettstown, NJ 07840
(performance analysis computer program)

Ungar Products Co., Inc.
Division of Eldon Industries
Buena Park, CA 90620
(soldering irons & accessories)

Publishers

A to Z Radio Control Electronics Journal
c/o George Steiner
2238 Rogue River Dr.
Sacramento, CA 96826

Electric Flight International
Traplet Publications Ltd.
Traplet House
Severn Drive, Upton-upon-Severn
Worcestershire WR8 0JL
United Kingdom

Electric Flight U.K.
c/o Dave Durnford
32 West Drayton Rd.
Hillingdon, Middlesex UB8 3LA
United Kingdom

Flying Models
Carstens Publications
P.O. Box 700
Newton, NJ 07860

Model Airplane News
Air Age Inc.
251 Danbury Rd.
Wilton, CT 06897

Model Aviation
Academy of Model Aeronautics (AMA)
5151 East Memorial Dr.
Muncie, IN 47302

Model Builder
Gallant Publications Inc.
34249 Camino Capistrano
Capistrano Beach, CA 92624

Quiet Flight International
Double Digit Publications Ltd.
5 Wootton Rise, Wootton Wawen
Warwickshire B95 6BJ
United Kingdom

RC Modeler
144 West Sierra Madre Blvd.
Sierra Madre, CA 91024

RC Reports
P.O. Box 1706
Huntsville, AL 35807

SAM Speaks
c/o Larry Clark
P.O. Box 528
Lucerne Valley, CA 92356

John Worth
newsletter on tiny R/C systems
4326 Andes Dr.
Fairfax, VA 22030

Index

a

A to Z Radio Control Electronics Journal, 59, 99
AC power supply, 40
Academy of Model Aeronautics (AMA), 62, 83, 99
ACE R/C Inc., 61, 93, 96
adjustable chargers, 35
Aerocraft Inc., 97
Aero Comp, 67
Aeronca L-3 Defender, 6
AI/Robotics, 47, 94
Airtronics Inc., 61, 96, 97
All Up-Last Down, 88
almost-ready-to-fly (ARF), 78
Altech Marketing, 73
AMA, (Academy of Model Aeronautics), 62, 83, 99
amp hours, 31
ampere hours, 31
amps, 14
analysis, 11
Anderson Power Pole, 50
Anthem Metal Products Inc., 95
ARF, (almost-ready-to-fly), 78
arming switch, 57
Astro Flight Inc., 3, 21, 39, 46, 53, 91, 93, 94, 95, 96,
Aveox Inc., 20, 91, 94

b

Balcomb, Phillip, 82
Baron, Dave, 85
Basic Flight Training, 82
battery chargers, 35
battery eliminator circuit (BEC), 42, 45, 55
battery heating, 34
battery testing device, 32
BEC, (battery eliminator circuit), 42, 45, 55
Begin-Aire, 77
belt drive, 79
belt reduction drive, 21, 66
blimps, 74
Boeing Model 314 PAA Clipper, 70
Boucher, Bob, 3
Boucher, Roland, 3
Bousquet, Don, 70
British Electric Flight Assoc. (BEFA), 86, 89
Brown, Dave, 83, 98
brushless DC motor, 17, 19, 20, 48
B&T R/C Products Inc., 93
Buholzei, 46
building, from scratch, 66
by-pass capacitor, 24

c

C/10 rate, 35
camcorder, 70
Cannon R/C Systems Inc., 61, 96
capacitor, 24
capacity, 32
center of gravity (c.g.), 83
Cermack Electronics & Model Supplies Co., 91, 93, 94
charging jacks, 57
cobalt, 17, 18, 19
CompuServe, 82
computer programs, 67
connectors, 49
Conquest 4NBF, 11
Controline, 72
convert, wet power, 66
cost, 12
crimping tool, 51
crow, 60
current carrying capability, 54

d

D.A.D. Inc., 62, 93, 96
Dave Brown Products Inc., 83, 98
deadstick, 84
Deans Company, 54, 96
Deans connector, 54
Defender, Aeronca L-3, 6
digital readout capabilities, 40
direct drive, 21, 64
dirigibles, 74
discharge testing, 40
drive, 63
Durnford, Dave, 86

e

"E" type cells, 39
Eclipse, 77
Electra, 77
Electri CUB 77
Electric Flight, 1
Electric Flight International, 85, 99
Electric Flight U.K., 86, 99
electric powered sailplane, 87
electrical brake circuit, 45
electrical noise, 23
Electro Flight Design Program, 67
Electromaster, 78
engine, 17
expanded scale voltmeter (ESV), 40

f

FAI series, 21
fan, 12-volt, 34
Federal Communications Commission (FCC), 62
ferrite, 11, 17, 20
flame-out, 84
flaperons, 60
Flightec Inc., 91, 94
Flying Models, 71, 77, 99
folding props, 26
frame rate, 42
frame rate controller, 42
freeflight, 71
fuse, 55
Futaba, 11, 59, 96
FX-35D, 47, 79

g

gauge, 21, 54
gear, 21
gear reduction drives, 64
Gibbs, Dick, 85
glass filled motor mount, 23
Graupner, 20

h

H-1000, 79
helicopters, electric, 71
high rate, 42
high rate controllers, 44
High Sky R/C Accessories, 41, 94
HiLine Ltd., 24, 91, 92, 93, 98
Hitec RCD Flash, 60
Hitec RCD Inc., 62, 93, 94, 96, 97
Hobby Lobby International, 91, 92, 93, 97
Horizon Hobby Distributors,62, 96
Horten flying wing, 72
Hunt, Tom, 72, 79

i

indoor, 73

j

Jaffe, Roger, 85
Jomar Products Inc., 42, 46, 54, 94

JR Remote Control, 62, 96
JR XF-622, 60

k

Kenway Micro Flight, 91, 95, 98
Keystone R/C Club, 87
kit, 66
Kopski, Bob, 75, 85
KRC Electric Fly, 87, 89
Kress, Bob, 67
Kress Jets, 67, 91, 92, 93, 97, 98
Kress, John, 67

l

Landing Products, 92
Leisure Electronics, 91, 95
Liberty 45, 78
Lightning Bug, 77
local instructor, 83
Lofty Pursuits Inc., 46, 94
low rate, 42
Lozier, Herb, 1, 3

m

Mabuchi, 11
Magnum, 33
Master Airscrew, 28
MaxCim Motors, 91, 94
MAX Packs, 33
micro computer control, 47
Militky, Fred, 3
Miller, Dick, 71, 85
milliamp-hours, 31
Model 110XL, 39
Model 112PK, 39
Model 4NBF conquest, Futaba, 59
Model 6 VA Sky Sport, 59
Model 7UGFS, Futaba, 60
Model Airplane News, 99
MODELAIR-TECH, 91, 95, 97, 98
Model Aviation, 76, 99
Model Builder, 99
Model Electronics Inc., 91, 95, 97
Modelnet, 82
motor, 17, 18, 19
motor run duration, 64
Mountjoy, Dr. John, 85

n

National Electric Aircraft Council (NEAC), 89
NEAC, Chairperson, 89
negative polarity, 15
New Creations R/C, 91, 94, 97
Ni-Cd, 30
Nickel-cadmium (Ni-Cd), 3
Nickel-metal hydride (NiMH), 37
NiMH (nickel-metal hydride), 37
nominal voltage, 30

o

Ogushwaitz, paul, 67
Old Timer events, 87
optical coupling, 46
overnight charge rate, 32, 35

p

Panasonic, 32
pattern, 69
peak detection charger, 38
Perfect Trainer (PT), 11
Pica-Robbe Inc., 91, 94, 95, 97
polarity, 15, 22
Poling, Mitch, 85
Polks Modelcraft Hobbies, 62, 96
positive polarity, 15
Primary Trainer Electric, 77
prop brake, 27

prop pitch, 27
propellers, 25
pylon racing, 69

q

Quiet Flight International, 85, 99

r

Radio Queen, 3
Radio Shack, 24, 52, 96, 98
RBRC, 37
R/C flying field, 81
RC Modeler, 78, 99
RC Reports, 99
R/C system components, 59
"ready to run", 90
receiver, 59
Rechargeable Battery Recycling Corp. (RBRC), 37
relay switches, 42
re-timing, 66
Revolt, 76
Robbe, 42

s

Safety inspection, 82
SAM Speaks, 99
Sandpiper, 78
Sanyo, 32
SCE, 33
Schmaedig Stick, 78
SCR, 37, 38
SCRC, 33, 80
Sermos, 57
Sermos connectors, 50
Sermos, John, 50
Sermos R/C Snap Connectors, 50, 96
Sermos Super Connector, 53
servo, S-133, 59
Shaw, Dr. Keith, 72, 85
Silentius, 3
simple switches, 41
simulation systems, 83
Skyvolt, 75
Slegers International, 46, 92, 95, 97
slow start, 27
Smart Charger/Cycler (SC/C), 40
Snappy Free Flight Timer, 72
Society of Antique Modelers (SAM), 87
sockets, 57
soft first flight, 36
software upgrades, 40
solder, 52
soldering iron, 51
solid state speed controls, 41
Sommerauer, 46
SonicTronics Inc., 23, 92
spoilerons, 60
SR, 32
SR Batteries Inc., 54, 87, 92, 93, 95, 96
Sribnick, Larry, 85, 89
standard Ni-Cd cell, 32
Steiner, George, 59
Super-Flex wire, 54
Super Seven, 60
Symposium on Electric Powered Flight, 87

t

Taplin, H.J, 3
Tarling, Gordon, 46, 72
theoretical motor run time, 37
thermistors, 40
TNR Technical Inc., 93
Today's Hobbies Inc., 76, 97
totally disabled, 57
Tower Hobbies Inc., 92 93, 96, 97

trainers, 69
training system, 83
trickle level, 38
turning phenomena, 84

u

Ultra Micro R/C, 61
Ungar Products Co. Inc., 51, 98
USR&D Inc., 67, 98

v

V-1 Buzz Bomb, 73
VL Products Inc., 92

w

watt, 13
watts, 20
wattage, 21
watts, 20
"wet" power, 7, 66
wicking, 52
Willard, Ken, 78
Windsor Propeller Co., 28, 92, 95
Winters, Bill, 77
wire, 49
Worth, John, 73, 99

y

E.F. Yost & Co., 93

z

zero-loss connector, 53